Advances in Quantum Electrodynamics Fundamentals

From the Electromagnetic Vacuum to Cosmology

Advances in Quantum Electrodynamics Fundamentals

From the Electromagnetic Vacuum to Cosmology

$$i\begin{pmatrix}\xi \\ \hbar\end{pmatrix}\frac{\partial}{\partial t}\left|\Phi_{k\lambda}\right\rangle = \begin{pmatrix}\tilde{\alpha}_{0k} \\ \tilde{H}_k\end{pmatrix}\left|\Phi_{k\lambda}\right\rangle$$

Constantin Meis

National Institute for Nuclear Science and Technology
CEA - Saclay, France

World Scientific

NEW JERSEY • LONDON • SINGAPORE • BEIJING • SHANGHAI • HONG KONG • TAIPEI • CHENNAI • TOKYO

Published by

World Scientific Publishing Co. Pte. Ltd.
5 Toh Tuck Link, Singapore 596224
USA office: 27 Warren Street, Suite 401-402, Hackensack, NJ 07601
UK office: 57 Shelton Street, Covent Garden, London WC2H 9HE

Library of Congress Control Number: 2026006459

British Library Cataloguing-in-Publication Data
A catalogue record for this book is available from the British Library.

ADVANCES IN QUANTUM ELECTRODYNAMICS FUNDAMENTALS
From the Electromagnetic Vacuum to Cosmology

ISBN 978-981-98-2665-0 (hardcover)
ISBN 978-981-98-2666-7 (ebook for institutions)
ISBN 978-981-98-2667-4 (ebook for individuals)

For any available supplementary material, please visit
https://www.worldscientific.com/worldscibooks/10.1142/14687#t=suppl

Desk Editor: Aanand Jayaraman

Typeset by Stallion Press
Email: enquiries@stallionpress.com

In memory of Hypatia
Alexandria, Egypt, 370–415 AD
"Use your mind to think, for even if you make mistakes it's better than not to think at all"

About the Author

Constantin Meis is Professor at the National Institute for Nuclear Science and Technology and Research Director at the CEA (Atomic Energy Commission, Saclay, France). He is also a member of the editorial board of the journal, *Radiation* (MDPI), Switzerland.

Professor Meis obtained "Maitrise de Physique", with specialization in atomic and nuclear physics, from University Paris 7 in 1984 and master's degree (DEA) in atomic physics and quantum optics from Orsay University (Paris-Sud) in 1985. He completed his PhD in atomic physics from Orsay University in 1988 and his HDR (Habilitation à Diriger de Recherches) from Marseille University in 2003.

Professor Meis started his career in 1988 as a physicist at the CEA Saclay, working on theoretical studies on computational modelling of the electromagnetic wave propagation and absorption in magnetized plasmas. In 2001, he was appointed as the Head of the Physics and Materials Unit of the National Institute for Nuclear Science and Technology, the academic department of the CEA, where he became Professor and then Research Director later. He published more than 100 papers and also authored the book *Light and Vacuum, 2nd Edition*, published by World Scientific in 2017. His main interests include the development of the quantum electrodynamics theory fundamentals as well as the understanding of the real nature of single photons.

Acknowledgements

Throughout the last years I had the opportunity and the honor to exchange on the quantum electrodynamics fundamentals with many renowned professors whose comments greatly helped me to advance in these studies. Hence, I'm very indebted to professor Pierre–Richard Dahoo (Versailles University, France), professor Christos Flytzanis (Ecole National Supérieure, France), professor Michael Raymer (Oregon University, USA), professor Evangelos Gazis (National Technical University, Greece) and professor Sergei Popov (KTH, Sweden).

I'm also particularly grateful to Professors Ahmed Hassanein (Purdue University, USA) and François Blanchard (Ecole de Technologies Supérieure, Canada) for the very fruitful discussions and suggestions.

Science is nothing but perception.

— Plato

Contents

Prologue

Quantum electrodynamics (QED) theory was developed during the 1930s–1960s in order to describe the quantum nature of light and its interaction with matter. Historically, the notion of the photon was introduced by Einstein based on Planck's studies on the radiation emitted by a black body. In his famous 1905 paper [1,2], he wrote as follows:

"The energy of a light ray spreading out from a point source is not continuously distributed over an increasing space but consists of a finite number of energy quanta, localized at points in space, which move without dividing and which can only be produced or absorbed as complete units."

The term "light quanta" was coined in this paper, while the term "photon" was introduced almost 20 years later by the chemist Lewis [3] using the Greek word $\phi\omega s$ (phos), which means light. In the same paper, Einstein advances an interpretation of the photoelectric effect, first observed by Hertz [4], employing a "point light quanta" model. A few years later, Compton used the same principles for the interpretation of the experimental results on X-ray scattering by electrons [5]. Thus, at the beginning of the twentieth century, the fundamental physical concept of the quantum nature of light was progressively accepted by the scientific community, and the QED theory began to be developed first by Dirac in 1927 [6] and then by many others in the 1930s and afterwards.

However, during the same period, in the beginning of the twentieth century, other studies carried out by Wentzel and Beck on the photoelectric effect, as well as those by Klein and Nishina on Compton's scattering, have demonstrated that these experiments can be interpreted quite well by using the electromagnetic wave nature of light without invoking the photon concept at all [7–9]. Hence, despite the continuous development of the QED theory in the decades after the 1930s, many renowned scientists still had the conviction that the photon concept is completely useless and that all the experiments can be interpreted by using the wave nature of light. For instance, Lamb was a fervent opposant to the photon concept and, with his colleague Scully, published an excellent interpretation of the photoelectric effect using the electromagnetic wave representation of light [10]. Even recently, in his well-known paper entitled "Anti-photon" [11] published in 1995, after analyzing extensively some selected experiments, he concluded as follows:

"It is high time to give up the use of the word 'photon', and of a bad concept which will shortly be a century old. Radiation does not consist of particles, and the classical, i.e., non-quantum, limit of QTR is described by Maxwell's equations for the electromagnetic fields, which do not involve particles."

In fact, various experiments have demonstrated that the photon should have a lateral expansion proportional to the quarter of its wavelength [12–15], while it was advanced that it is impossible to localize a photon in a length shorter than roughly its wavelength [16]. Indeed, experiments carried out by Mandel and others have concluded that it is impossible to localize a photon better than in a volume corresponding roughly to the cube of its wavelength [17,18]. Consequently, the photon concept initially advanced by Einstein, as reported above, does not correspond to a real physical state since the experimental evidence has shown that photons cannot be points and they are not absorbed as a whole because they can interact with matter, conceding a part of their energy by changing frequency, e.g., Compton's scattering. This statement was also used by Lamb, advancing that "*photons cannot be localized in any meaningful manner, and they do not behave at all like particles, whether described by a wave function or not.*"

Despite all that, the point photon model has been used for the development of the QED theory. This was mainly due to the fact that the point particle concept, although it is naturally unable to provide a satisfactory description of a single photon, is an extremely practical and efficient representation in order to describe states before and after an interaction process. For instance, the interpretation of the photoelectric effect based on the point photon model is a simple equation, while that based on the electromagnetic waves extends over roughly a page of calculation. However, the last one is considered by many scientists to be more realistic and precise.

Obviously, the continuous conflictual positions among many pioneers of modern physics upon the wave or particle nature of light is due to the fact that it was never possible and we are still unable today to describe with a satisfactory mathematical representation a single photon whose all individual properties are compatible with the experimental observations. Even Bohr's *complementarity principle*, announced apparently with the purpose of satisfying both opponents, was ruled out by the double prism experiment, which validated that a photon can demonstrate both the particle and wave natures in the same experiment [19].

In recent years, the development of parametric down-conversion techniques has permitted us to realize experiments with single photons, complicating even more our ideas on the nature of the photon [20,21]. Thus, Grangier *et al.* demonstrated experimentally the indivisibility of photons [22], while the entangled state experiments [23] have indicated that the photon should have a real non-local wave function, although the detectors react to a localized single particle energy.

Resuming the overall experimental evidence, the photon appears to be a *wave–corpuscle* in which the electromagnetic field preserves the wave form over at least a wavelength, with circular polarization corresponding to the spin. The whole *wave–corpuscle* has well-defined energy and momentum and propagates guided by a non-local wave function extending to infinity. This is certainly a partial picture composed of the principal physical properties deduced from the experiments up to today, but it provides an idea of the difficulties faced in developing a concrete mathematical representation of the photon. However, the quantization of the electromagnetic field in QED is based on a simple harmonic oscillator formalism, although no experiments have ever confirmed that a single photon state is a harmonic oscillator. Also, the harmonic oscillator quantization procedure in the case of the electromagnetic field involves a well-known fundamental mathematical ambiguity between the canonical variables of position and momentum and the corresponding quantum mechanics Hermitian operators [24–26]. Hence, the harmonic oscillator formalism cannot really represent the quantum state of the electromagnetic field. This is also the reason why, in QED, the normal ordering Hamiltonian is mainly used, casting aside the harmonic oscillator representation.

Furthermore, the principal singularity in QED is due to the zero-point energy of the harmonic oscillator, whose energy density is infinite at any point in space. Even by introducing a reasonable cut-off, corresponding to Planck's energy, the calculated zero-point energy density is roughly 120 orders of magnitude higher than that estimated by recent well-validated astrophysical observations. This huge discrepancy, usually called the quantum vacuum catastrophe, constitutes a fundamental riddle in modern physics [27, 28]. Therein, it is also extremely important to underline that the vacuum effects, like spontaneous emission and Lamb shifts, usually attributed to the zero-point energy, are in fact calculated in QED based on the commutation properties of the photon creation and annihilation operators without invoking the zero-point energy term at all [24–26, 29]. As for the Casimir effect, also attributed usually to the zero-point state of the electromagnetic field, it has been demonstrated by many studies using different methods that it is due to space charge distributions and Lorentz's forces and can be calculated exactly without referring to the zero-point energy [30–33]. Following the above, the zero-point energy of the electromagnetic field issued from the harmonic oscillator Hamiltonian cannot represent a real physical state.

Obviously, despite the extremely fast experimental evolution in the last decades, the QED theory has not evolved accordingly in order to take into account the quite consistent accumulation of new results regarding the drawbacks related to the point photon concept, the photon wave function and the harmonic oscillator representation, including the zero-point singularity. Thus, we present here a synthesis of successively published studies with the aim of gaining a clearer idea of the photon structure compatible with the experimental evidence. We address the essential and fundamental questions of QED pertaining to the electromagnetic field quantization, the photon vector potential, the photon wave function and finally the electromagnetic vacuum representation independent of the zero-point energy.

Finally, we show that the direct relationship between the photon and the electron–positron charge as well as the electromagnetic vacuum, a zero-energy field permeating all space, issue naturally from the vector potential quantization. The electromagnetic vacuum is at the origin of the observed vacuum effects. The energy density corresponding to its spontaneous fluctuations is compatible with the astrophysical observations and is related to the cosmological constant Λ. It is concluded that the gravitational constant G depends on the electromagnetic vacuum density of states and might not be universal. In addition, the masses of all elementary particles appear to come from the electromagnetic vacuum states. Thus, without stating postulates or advancing any assumptions, the quantization of the electromagnetic field vector potential naturally links QED to particle physics, string theory and cosmology.

References

[1] A. Einstein, Concerning an heuristic point of view toward the emission and transformation of light, *Ann. Phys.* 322, 132 (1905).

[2] A. B. Arons and M. B. Peppard, Einstein's proposal of the photon concept—A translation of the Annalen der Physik paper of 1905, *Am. J. Phys.* 33, 367 (1965).

[3] G. N. Lewis, The conservation of photons, *Nature* 118, 874 (1926).

[4] H. Hertz, Ueber sehr schnelle electrische Schwingungen, *Ann. Phys.* 267, 983 (1887).

[5] A. H. Compton, A quantum theory of the scattering of X-rays by light elements, *Phys. Rev.* 21, 483 (1923).

[6] P. A. M. Dirac, The quantum theory of the emission and absorption of radiation, *Proc. R. Soc. London A.* 114(767), 243 (1927).

[7] G. Wentzel, Zur Theorie des photoelektrischen Effekts, *Z. Phys.* 40, 574 (1926).

[8] G. Beck, Zur Theorie des Photoeffekts, *Z. Phys.* 41, 443 (1927).

[9] O. Klein and Y. Nishina, Über die Streuung von Strahlung durch freie Elektronen nach der neuen relativistischen Quantenmechanik, *Z. Phys.* 52, 853 (1929).

[10] W. E. Lamb, Jr. and M. O. Scully, in *Polarisation, Matière et Rayonnement, Volume Jubilaire en l'Honneur d'Alfred Kastler* (French Physical Society, Presses Universitaires de France, Paris, France, 1969).

[11] W. E. Lamb, Jr., Anti-photon, *Appl. Phys. B.* 60, 77–84 (1995).

[12] H. L. Robinson, Diffraction patterns in circular apertures less than one wavelength in diameter, *J. Appl. Phys.* 24(1), 35 (1953).

[13] R. K. Hadlock, Diffraction patterns at the plane of a slit in a reflecting screen, *J. Appl. Phys.* 29, 918 (1958).
[14] G. Hunter and R. L. P. Wadlinger, Photons and neutrinos as electromagnetic solitons, *Phys. Essays* 2, 158 (1989).
[15] G. Hunter, in *Quantum Uncertainties: Recent and Future Experiments and Interpretations*, W. M. Honig, D. W. Kraft, and E. Panarella, Eds., NATO ASI Series (Series B: Physics), Vol. 162 (Springer, Boston, MA, USA, 1987), p. 331.
[16] A. I. Akhiezer and V. B. Berestetskii, *Quantum Electrodynamics* (Interscience, New York, NY, USA, 1965).
[17] L. Mandel, Configuration-space photon number operators in quantum optics, *Phys. Rev.* 144, 1071 (1966).
[18] R. L. Pfleegor and L. Mandel, Interference of independent photon beams, *Phys. Rev.* 159(5), 1084 (1967).
[19] P. Ghose and D. Home, The two-prism experiment and wave-particle duality of light, *Found. Phys.* 26(7), 943 (1996).
[20] D. N. Klyshko, A. N. Penin, and B. F. Polkovnikov, Parametric luminescence and light scattering by polaritons, *JETP Lett.* 11, 5 (1970).
[21] D. C. Burnham and D. L. Weinberg, Observation of simultaneity in parametric production of optical photon pairs, *Phys. Rev. Lett.* 25(2), 84 (1970).
[22] P. Grangier, G. Roger, and A. Aspect, Experimental evidence for a photon anticorrelation effect on a beam splitter: A new light on single-photon interferences, *Europhys. Lett.* 1(4), 173 (1986).
[23] G. Auletta, *Foundations and Interpretation of Quantum Mechanics* (World Scientific, Singapore, 2001).
[24] J. C. Garrison and R. Y. Chiao, *Quantum Optics* (Oxford University Press, Oxford, UK, 2008).
[25] G. Grynberg, A. Aspect, and C. Fabre, *Introduction to Quantum Optics* (Cambridge University Press, Cambridge, UK, 2010).
[26] C. Meis, Quantized field of single photons, in *Single Photon Manipulation*, K. Xia, Ed. (IntechOpen, London, UK, 2020).
[27] T. Padmanabhan, Cosmological constant: The weight of the vacuum, *Phys. Rep.* 380, 235 (2003).
[28] J. A. Frieman, M. S. Turner, and D. Huterer, Dark energy and the accelerating universe, *Annu. Rev. Astron. Astrophys.* 46, 385–432 (2008).
[29] C. Meis, *Light and Vacuum*, 2nd ed. (World Scientific, Singapore, 2017).
[30] P. W. Milonni, Casimir Forces without the Vacuum Radiation, *Phys. Rev. A* 25, 1315 (1982).

[31] H. Ezawa, K. Nakamura, and K. Watanabe, The Casimir Force from Lorentz's, in *Frontiers in Quantum Physics*, S. C. Lim, R. Abd-Shukor, and K. H. Kwek, Eds. (Springer, Singapore, 1998), pp. 160–169.
[32] G. Gründler, The Casimir effect: No manifestation of zero-point energy, arXiv:1303.3790 (2013).
[33] J. Schwinger, L. L. DeRaad, Jr., and K. A. Milton, Casimir effect in dielectrics, *Ann. Phys.* 115, 1 (1978).

Part I

Electromagnetic Field Quantization, Photon Structure and Wave Function, the Electromagnetic Vacuum

Introduction

In this part, the following fundamental concepts are investigated and analyzed:

- photon vector potential quantization independent of any external cavity parameter;
- quantization constant of the photon vector potential;
- photon vector potential amplitude operator;
- relation between the electron–positron charge and the photon vector potential;
- electromagnetic vacuum overcoming the zero-point energy singularity;
- relation of the photons, electrons and positrons to the electromagnetic vacuum;
- photon electric and magnetic fields independent of any external volume parameter;
- photon quantum equation, representing both the first and second quantization of the electromagnetic field;
- photon wave function;
- relation between the photon wave function and the electromagnetic vacuum;
- photon quantization volume.

Chapter 1

Enhanced Vector Potential Quantization and Photon Wave–Particle Representation*

Abstract
The relationship between the photon vector potential amplitude and its frequency is analyzed. The analogy between the relativistic quantum mechanical equations for a massless particle and those governing the photon vector potential appears clearly. Finally, the virtual electromagnetic waves associated with the photon and predicted by de Broglie, Bohr and others appear naturally as a result of the photon vector potential quantization. The electromagnetic vacuum state is directly deducted from the quantized photon vector potential.

1. Vector potential quantization enhanced at a single photon

The wave–particle nature of light as well as the vacuum electrodynamical entity are no doubt among the most peculiar notions introduced by modern physics. The well-known Planck–Einstein relations $\{E = \hbar\omega; \vec{p} = \hbar\vec{k}\}$, where $\hbar$ is Planck's reduced constant, associated with the electromagnetic wave angular frequency ω and wave vector $\vec{k}$ to the photon energy E and momentum $\vec{p}$. Louis de Broglie

*This chapter is reproduced from C. Meis, 1997. "Photon wave-particle duality and virtual electromagnetic waves", *Foundation of Physics*, 27(6), pp. 865–873.

extended these relations to particles. Despite the multitude of experimental confirmations of the photon duality, the concept of a localized particle carrying all the energy and the momentum and at the same time behaving as a wave was (and continues to be) hardly understood. Bohr introduced the idea of "virtual waves" [1] equivalent to Lorentz's "ghost waves" [2] in order to clarify the wave–particle nature of the light. Of course, it is well known that, for Bohr and Lorentz, the notion of light's dualism was not exactly the same, but we can essentially say that, for both of them, a wave with zero energy and momentum has to propagate in space with the particle inducing quantum transitions in matter. Louis de Broglie also had similar ideas, with a point of view rather closer to Lorentz's concept [3, 4] and proposed his remarkable double solution theory. Nevertheless, the majority of physicists were extremely skeptical about the physical reality of any description based upon waves without energy and momentum. Since the efforts for a perfect mathematical representation of electromagnetic waves and photon unification remained fruitless, Bohr announced the famous complementarity principle [5, 6]. In later years, the quantum electrodynamics theory [7], in order to explain the spontaneous emission of photons by excited atoms, introduced the notion of a dynamical vacuum consisting of the tensor product of all k-mode and λ-polarization zero photon occupancies $|n_{k\lambda} = 0, \ldots, n_{k'\lambda'} = 0\rangle$, which is the fundamental eigenstate of a harmonic oscillator Hamiltonian [8]. In this representation, the vacuum proves to be an entity of infinite energy $\sum_{k,\lambda} \frac{1}{2}\hbar\omega_k$ and infinite field intensity whose fluctuations can induce quantum transitions. Renormalization operations had to be introduced in order to circumvent the infinite energy difficulty [9, 10]. Consequently, we can say that the actual scientific picture on the nature of light is not completely satisfactory and still puzzles a lot of physicists [1, 12, 13].

Energy and momentum imply particle properties, while frequency and vector potential imply electromagnetic wave properties. It is impossible to define completely an electromagnetic wave only by the frequency, as there is no information about the energy density for the definition of which the vector potential is necessary. Herein, it is worth noting that the Planck–Einstein relations attribute a precise value of energy and momentum to a single photon related to the electromagnetic wave only by the intermediate of the frequency while giving no information about its vector potential. From a physical point of view, this is rather due to the fact that the quantum representation of the electromagnetic wave is not characterized by a natural constant relating the vector potential to the frequency, analogous to Planck's constant associated with the energy quantization. In what follows, the possibility of the existence of such a relation is analyzed.

The energy density of a transverse electromagnetic wave in vacuum is given, in a classical description, by the well-known expression [14]

$$W_k(\vec{r},t) = \frac{1}{2}\left(\varepsilon_0|\vec{E}_k(\vec{r},t)|^2 + \frac{1}{\mu_0}|\vec{B}_k(\vec{r},t)|^2\right) \tag{1}$$

where $\vec{E}_k(\vec{r},t)$ and $\vec{B}_k(\vec{r},t)$ are the electric and magnetic fields respectively of the k-mode wave, with ε_0 and μ_0 the electric permittivity and magnetic permeability of the vacuum, respectively, satisfying the relation $\varepsilon_0\mu_0 c^2 = 1$, where c is the speed of light in vacuum.

Recalling that the electric and magnetic fields are proportional to the vector potential $\vec{A}_k(\vec{r},t)$ following the expressions $\vec{E}_k(\vec{r},t) = \omega_k\vec{A}_k(\vec{r},t)$ and $c\vec{B}_k(\vec{r},t) = \omega_k\vec{A}_k(\vec{r},t)$, the energy density writes as a function of the vector potential

$$W_k(\vec{r},t) = \varepsilon_0\omega_k^2|\vec{A}_k(\vec{r},t)|^2 \tag{2}$$

Now, a photon considered as an integral particle with a given energy, momentum and frequency should also have a definite value of a vector potential. Suppose it is possible to reduce Eq. (2) in the case of a single photon of energy $\hbar\omega_k$. Then one should have

$$\int \varepsilon_0\omega_k^2\alpha_{0k}^2(\omega_k)\left[\hat{\varepsilon}_{k\lambda}e^{i(\vec{k}\cdot\vec{r}-\omega_k t+\theta)} + \mathrm{cc}\right]^2 d^3r = \hbar\omega_k \tag{3}$$

where cc denotes the complex conjugate and

$$\vec{\alpha}_k(\omega_k,\vec{r},t) = \alpha_{0k}(\omega_k)\left[\hat{\varepsilon}_{k\lambda}e^{i(\vec{k}\cdot\vec{r}-\omega_k t+\theta)} + \mathrm{cc}\right] \tag{4}$$

is a wave function assumed to be that of the photon vector potential satisfying the wave equation

$$\frac{\partial^2}{\partial t^2}\vec{\alpha}_k(\omega_k,\vec{r},t) - c^2\vec{\nabla}^2\vec{\alpha}_k(\omega_k,\vec{r},t) = 0 \tag{5}$$

On the other hand, in order to respect the photon particle entity in the sense of Einstein's concept [15, 16], Eq. (3) must hold at any instant t. Consequently, the square of the vector potential must be time-independent. For this, the polarization unit vector $\hat{\varepsilon}_{k\lambda}$ must have at least two orthogonal components, $\hat{e}_1$ and $\hat{e}_2$, such as $\hat{\varepsilon}_{k\lambda} = a_{k\lambda}^{(1)}\hat{e}_1 + a_{k\lambda}^{(2)}\hat{e}_2$ with $|a_{k\lambda}^{(1)}|^2 + |a_{k\lambda}^{(2)}|^2 = 1$ and $\hat{e}_1 \cdot \hat{e}_2 = 0$. For the sake of simplicity and in order to satisfy the above conditions, a circular polarization will be considered here.

Thus, $\hat{\varepsilon}_{k\lambda} = \frac{1}{\sqrt{2}}(\hat{e}_1 \pm i\hat{e}_2)$, so Eq. (3) becomes

$$\int 2\varepsilon_0\omega_k^2\alpha_{0k}^2(\omega_k)d^3r = \hbar\omega_k \tag{6}$$

It is impossible to integrate the last relation without knowing explicitly the local spatial distribution of the photon in which the vector potential amplitude is constant. As the square of $\alpha_{0k}(\omega_k)$ is just a scalar quantity defined on a point of the propagation axis, Eq. (6) seems to be the integration of a "ghost" entity. Nevertheless, we can use different physical features in order to circumvent this difficulty. The directional character of photons has been demonstrated experimentally by Compton [17]. In the classical electromagnetic field description resulting from Maxwell's equations [14], the vector potential precesses (rotates) perpendicular to the propagation axis, while the amplitude $\alpha_{0k}(\omega_k)$ of a circular polarization is fixed for a given value of frequency and remains constant during the propagation. Therefore, the directional sense has to be involved in Eq. (6) by integrating in cylindrical coordinates (r, θ, z), where z is the propagation axis:

$$\int_0^{\infty}\int_0^{2\pi}\int_{-\infty}^{+\infty} 2\varepsilon_0\omega_k^2\alpha_{0k}^2(\omega_k)r dz d\theta dr = \hbar\omega_k \tag{7}$$

It has been demonstrated that a photon of a given energy, and hence of a given frequency, cannot be considered as an integral particle within a region smaller in order of magnitude than a wavelength λ_k [7]. Consequently, Eq. (7) must be integrated in the interval $0 < z < \lambda_k$ in which of course θ and z are not independent variables, as $\theta = 2\pi z/\lambda_k$. Thus, the integration can be reduced only to the variables z and r. Now, the fact that at any point of the propagation axis, the vector potential amplitude has a definite constant value depending only on the angular frequency is quite important because $\alpha_{0k}(\omega_k)$ behaves as if it "knows" its angular frequency even in the "beginning" of the "first" precession. Thus, we can assume that the space in which $\alpha_{0k}(\omega_k)$ is constant can be expressed in such a way that $\alpha_{0k}(\omega_k) = \alpha_{0k}\Lambda(z)\Pi(r)$, where $\Lambda(z)$ and $\Pi(r)$ are functions, depending only on z and r, respectively. Hence, we introduce the function $\Lambda_{\lambda_k}(z)$ defined as $\Lambda_{\lambda_k}(z) = 1$ for $0 \leq z \leq \lambda_k$ and $\Lambda_{\lambda_k}(z) = 0$ elsewhere. One gets

$$2\varepsilon_0\omega_k^2\alpha_{0k}^2(\omega_k)\int_0^{\infty}\int_{-\infty}^{+\infty}\Lambda_{\lambda_k}^2(z)\Pi^2(r)rdzdr = \hbar\omega_k \tag{8}$$

and thus,

$$2\varepsilon_0\omega_k^2\alpha_{0k}^2(\omega_k)\lambda_k\int_{\circ}^{\infty}\Pi^2(r)rdr = \hbar\omega_k \tag{9}$$

According to a dimension analysis, the vector potential amplitude should be proportional to the frequency so that we can put $\alpha_{0k}(\omega_k) = \xi\omega_k$, where ξ is a constant, so Eq. (9) becomes

$$2\varepsilon_0\xi^2\omega_k^4\lambda_k\int_{\circ}^{\infty}\Pi^2(r)rdr = \hbar\omega_k \tag{10}$$

Now, the left-hand side of (10) has to be homogeneous dimensionally to the right-hand side. Therefore, the integration should give a positive quantity proportional to λ_k^2. This leads to the interesting result that the radial distribution of $\alpha_{0k}(\omega_k)$ also depends on the wavelength. It could be of the form $\Pi(r) \propto e^{-pr/\lambda_k}$, where p is a constant. But if that is really the case, then the dispersion relation $\omega_k^2 = |\vec{k}|^2c^2$, used in Planck–Einstein relations, cannot be obtained by the propagation equation (5) unless $p = 0$. Thus, the radial function $\Pi(r)$ must be of the form $\Pi_{\lambda_k}(r) = 1$ for $0 \leq r \leq \eta\lambda_k$, and $\Pi_{\lambda_k}(r) = 0$ for $r > \eta\lambda_k$, with η a positive dimensionless constant. Equation (10) gives

$$2\varepsilon_0\xi^2\omega_k^4\left(\frac{\eta^2}{2}\lambda_k^3\right) = \hbar\omega_k \tag{11}$$

Comparison of the last equation with relation (6) shows that not only the photon vector potential is related to the frequency by a constant, but the "volume" of the photon should also have a definite value proportional to λ_k^3. The product of the two constants has the value

$$\eta\xi = \frac{1}{(2\pi)^{3/2}}\sqrt{\frac{\hbar}{\varepsilon_0 c^3}} \tag{12}$$

An equivalent result can be obtained by integrating in spherical coordinates for $0 \leq r \leq \lambda_k/2$ and considering that $\alpha_{0k}(\omega_k)$ is space-independent:

$$2\varepsilon_0\chi^2\xi^2\omega_k^4 \int_0^{\lambda_k/2} 4\pi r^2 dr = \hbar\omega_k \tag{13}$$

where χ^2 is a positive constant for the last equation to hold.

We get

$$\chi\xi = \frac{1}{(2\pi)^{3/2}}\sqrt{\frac{3\hbar}{\pi\varepsilon_0 c^3}} \tag{14}$$

The final impression of this calculation is that integrating an "empty" volume whose dimensions are of the order of the wavelength, a constant relating the photon vector potential amplitude to the angular frequency appears, whatever the type of the coordinates used. Perhaps, appropriate experiments could give a precise value to this constant.

2. Quantum equation governing the photon vector potential

We can analyze now what happens if such a constant exists. According to the previous results, the Planck–Einstein relations have to be completed as follows:

$$\frac{\alpha_{0k}}{|\xi|} = \frac{E_k}{\hbar} = \frac{|\vec{p}|c}{\hbar} = |\vec{k}|c = \omega_k \tag{15a}$$

where

$$\xi \propto \frac{1}{(2\pi)^{3/2}}\sqrt{\frac{\hbar}{\varepsilon_0 c^3}} \tag{15b}$$

The last expression relates the dispersion relation of an electromagnetic wave obtained from Maxwell's equations to the energy, momentum and vector potential amplitude of the photon considered as an integral particle. Uncertainty between the photon vector potential amplitude and time results from Heisenberg's principle and (15)

$$\delta E_k \delta t > \hbar \rightarrow \delta \alpha_{0k} \delta t > \xi \tag{16}$$

Now, $\vec{\alpha}_k(\omega_k, \vec{r}, t)$ should satisfy the wave propagation equation. Combination of Eq. (5) and Eq. (15) leads to the relation

$$[\alpha_{0k}^2(\omega_k) + \xi^2 c^2 \vec{\nabla}^2]\vec{\alpha}_k(\omega_k, \vec{r}, t) = 0 \tag{17}$$

Therefore, the photon vector potential amplitude can be expressed as an operator:

$$\tilde{\alpha}_{0k}(\omega_k) = -i\xi \frac{\omega_k}{|\vec{k}|} \vec{\nabla}_k = -i\xi c \vec{\nabla}_k \tag{18}$$

where the subscript k denotes simply that the gradient acts upon the mode k.

It is worth noting the symmetry with the relativistic Hamiltonian for a massless particle

$$\tilde{H} = -i\hbar c \vec{\nabla} \tag{19}$$

Applying now $\tilde{\alpha}_{0k}(\omega_k)$ to the photon vector potential $\vec{\alpha}_k(\omega_k, \vec{r}, t)$, we get

$$\tilde{\alpha}_{0k}(\omega_k)\vec{\alpha}_k(\omega_k, \vec{r}, t) = i\xi c|\vec{k}|\vec{\alpha}_k \left(\omega_k, \vec{r}, t, \theta + \frac{\pi}{2}\right) \tag{20}$$

From the other hand, the time variation of $\vec{\alpha}_k(\omega_k, \vec{r}, t)$ is written as

$$\frac{\partial}{\partial t}\vec{\alpha}_k(\omega_k, \vec{r}, t) = \omega_k \vec{\alpha}_k \left(\omega_k, \vec{r}, t, \theta + \frac{\pi}{2}\right) \tag{21}$$

From the last two relations, we obtain the equation governing the photon vector potential

$$i\xi \frac{\partial}{\partial t}\vec{\alpha}_k(\omega_k, \vec{r}, t) = \tilde{\alpha}_{0k}(\omega_k)\vec{\alpha}_k(\omega_k, \vec{r}, t) \tag{22}$$

The constant ξ is for the vector potential the equivalent of $\hbar$ for the energy. Consequently, the photon vector potential function $\vec{\alpha}_k(\omega_k, \vec{r}, t)$ satisfies a wave propagation equation (5) and a linear time-dependent quantum equation (22) whose quantization constant is ξ. Equations (16), (18), and (22) demonstrate a symmetry among the relativistic quantum mechanical equations for the energy of a massless particle and those governing the photon vector potential.

3. Photons from the electromagnetic vacuum

The photon vector potential function can be expressed as follows:

$$\vec{\alpha}_k(\omega_k, \vec{r}, t) = \omega_k \vec{\Xi}_k(\omega_k, \vec{r}, t) \tag{23}$$

where the principal function $\vec{\Xi}_k(\omega_k, \vec{r}, t)$ is

$$\vec{\Xi}_k(\omega_k, \vec{r}, t) = \xi \left(\hat{\varepsilon}_{k\lambda} e^{i(\vec{k}\cdot\vec{r} - \omega_k t + \theta)} + \hat{\varepsilon}^*_{k\lambda} e^{-i(\vec{k}\cdot\vec{r} - \omega_k t + \theta)} \right) \tag{24}$$

which also is written as an operator by using the k-mode and λ-polarization photon creation and annihilation non-Hermitian operators, $a^+_{k\lambda}$ and $a_{k\lambda}$, respectively:

$$\tilde{\Xi}_k(\omega_k, \vec{r}, t) = \xi \left(\hat{\varepsilon}_{k\lambda} a_{k\lambda} e^{i(\vec{k}\cdot\vec{r} - \omega_k t + \theta)} + \hat{\varepsilon}^*_{k\lambda} a^+_{k\lambda} e^{-i(\vec{k}\cdot\vec{r} - \omega_k t + \theta)} \right) \tag{25}$$

It is clear that $\vec{\Xi}_k(\omega_k, \vec{r}, t)$ is not a vector potential. $\vec{\Xi}_k(\omega_k, \vec{r}, t)$ is a vector field with amplitude ξ, carrying neither energy nor momentum and satisfying the wave propagation equation. It is worth noticing that this kind of "virtual waves" appear naturally when a quantification constant for the photon vector potential is considered. Furthermore, Eq. (25) shows that $\tilde{\Xi}_k(\omega_k, \vec{r}, t)$ can induce quantum electronic transitions. Another interesting result is the behavior of the field $\tilde{\Xi}_k(\omega_k, \vec{r}, t)$ when the angular frequency tends to zero which does not vanish and reduces to $\tilde{\Xi}_{0\lambda}(\omega_k = 0, \vec{r}, t) = \xi(\hat{\varepsilon}_{k\lambda} a_{k\lambda} e^{i\theta} + \hat{\varepsilon}^*_{k\lambda} a^+_{k\lambda} e^{-i\theta}))$. Thus, in absence of energy, momentum, and even angular frequency $\Xi_{0\lambda}(\omega_k = 0, \vec{r}, t)$ still subsists, being always capable of inducing quantum transitions. Under these conditions $\tilde{\Xi}_{0\lambda}(\omega_k = 0, \vec{r}, t)$ can only be associated to a vacuum field whose states constitute the "skeleton" of photons. $\tilde{\Xi}_{0\lambda}(\omega_k = 0, \vec{r}, t)$ can be considered responsible for the spontaneous emission. During the last processes the energy difference δE_{if} between the initial $|i\rangle$ and final $|f\rangle$ quantum states determines the angular frequency $\omega_{if} = \delta E_{if}/\hbar$ which is "conferred" to the vacuum state, generating directly a "virtual" wave $\vec{\Xi}_{if}(\omega_{if}, \vec{r}, t)$ and so a photon vector potential $\vec{\alpha}_{if}(\omega_{if}, \vec{r}, t) = \omega_{if} \vec{\Xi}_{if}(\omega_{if}, \vec{r}, t)$ with an amplitude $\alpha_{0if}(\omega_{if}) = \xi\omega_{if}$. The stimulated emission involves the vacuum state $\vec{\Xi}_k(\omega_k, \vec{r}, t)$ on the condition the angular frequency corresponds to the energy difference between the two quantum states.

According to Eq. (4) and (11), the classical wave properties are reduced within a volume of the order of λ_k^3 resulting in physical wave properties for individual photons expressed by the vector potential which behaves here as a photon wave function for Eq. (22). Furthermore, Eq. (23)–(25) show precisely that, in this representation, the wave properties of an individual photon derive from (and are strongly related to) a "virtual wave" corresponding to a vacuum state. The final picture one gets is that, following Eq. (15), (23) and (24), everything happens as if the only crucial fact for the "existence" of a photon, characterized by both particle and wave properties, is the angular frequency. Energy, momentum, wave vector, vector potential amplitude and even the virtual wave of the vacuum field $\vec{\Xi}_k(\omega_k, \vec{r}, t)$, which the photon consists of, are only defined for a non-zero value of the angular frequency. But the angular frequency of what? That of the vacuum quantized vector field $\vec{\Xi}_{0\lambda}(\omega_k = 0, \vec{r}, t)$ with amplitude ξ, whose precession at the angular frequency ω_k gives birth to a virtual wave $\vec{\Xi}_k(\omega_k, \vec{r}, t)$, from which results a real wave vector potential $\vec{\alpha}_k(\omega_k, \vec{r}, t) = \omega_k \vec{\Xi}_k(\omega_k, \vec{r}, t)$ with amplitude $\xi\omega_k$ and thus a quantum of energy $\hbar\omega_k$ and momentum $\hbar\vec{k}$.

The above description expresses mathematically what exactly P.A.M. Dirac had developed qualitatively in his famous 1927 paper about the Quantum Theory of the Emission and Absorption of Radiation [18]:

> *"The light-quantum has the peculiarity that it apparently ceases to exist when it is in one of its stationary states, namely, the zero state, in which its momentum, and therefore also its energy, are zero. When a light-quantum is absorbed it can be considered to jump into these zero states, and when one is emitted it can be considered to jump from the zero state to one in which it is in evidence, so that it appears to have been created."*

Finally, this representation is compatible with the experimental evidence [19, 20] following which the photon can only be localized within a volume of the order of the cube of its wavelength.

4. Conclusion

The existence of a quantization constant ξ relating the photon vector potential amplitude to its angular frequency ω_k cannot be excluded and has to be investigated experimentally. If that is really the case, the Planck–Einstein relations should then be completed in order to take into account the quantized photon vector potential which satisfies simultaneously a classical wave propagation equation and a Schrödinger-like quantum equation. In this way, the photon will be fully related to the electromagnetic wave, giving evidence of the wave–particle nature of light. The particle properties of the photon are derived from the energy and the momentum, while the wave properties result from the quantized vector potential field oscillating at the angular frequency ω_k. In this description, the vacuum appears naturally to be a field without energy but capable of interacting with the quantum energy states of matter. It comes out that the photon is an oscillation of a vacuum state. It is composed of a "virtual wave" $\vec{\Xi}_k(\omega_k, \vec{r}, t)$ without energy and momentum similar to that predicted by Bohr, Dirac and Lorentz.

References

[1] N. Bohr, H. A. Kramers, and J. S. Slater, The quantum theory of radiation, *Philos. Mag.* 47, 785 (1924).
[2] H. A. Lorentz, *Problems of Modern Physics* (Caltech Lectures of 1922) (Ginn, Boston, 1927; Dover, New York, 1967).
[3] L. de Broglie, *Une tentative d'interprétation causale et non-linéaire de la mécanique ondulatoire. La théorie de la double solution* (Gauthier-Villars, Paris, 1956).
[4] L. de Broglie, *La réinterprétation de la mécanique ondulatoire* (Gauthier-Villars, Paris, 1971).
[5] N. Bohr, Discussion with Einstein on Epistemological Problems in Atomic Physics, in A. Einstein: Philosopher-Scientist, P. A. Schilpp, ed. (Harper & Row, New York 1959).
[6] M. Jammer, *The Philosophy of Quantum Mechanics* (Wiley-Interscience, New York, 1974).
[7] A. I. Akhiezer and B. V. Berestetskii, *Quantum Electrodynamics* (Interscience, New York, 1965, p. 11).
[8] H. Haken, *Light* (North-Holland, Amsterdam, 1981).

[9] W. Heitler, *The Quantum Theory of Radiation* (Clarendon, Oxford, 1954).
[10] M. H Mittleman, *Theory of Laser Atom Interactions* (Plenum. New York, 1982).
[11] A. Szczepanski, State vector reduction and photon coincidences, *Found. Phys.* 6, 427 (1976).
[12] Mioara Mugur-Schfichter, Locality, reflection, and wave-particle duality, *Found. Phys.* 17, 813 (1987).
[13] E. Selleri, On the direct observability of quantum waves, *Found. Phys.* 12, 1087 (1982).
[14] J. D. Jackson, *Classical Electrodynamics*, 2nd edn (Wiley, New York, 1975).
[15] A. Einstein, Concerning an heuristic point of view toward the emission and transformation of light, *Ann. Phys.* (Bern) 17, 132 (1905).
[16] A. Einstein, The quantum theory of radiation, *Phys. Z.* 18, 121 (1917).
[17] A. H. Compton, A quantum theory of the scattering of X-rays and Y-rays by light elements, *Phys. Rev.* 21, 483 (1923).
[18] P. A. M. Dirac, The quantum theory of the emission and absorption of radiation, *Proc. Roy. Soc.*, A114, 243 (1927).
[19] M. Dagenais and L. Mandel, Investigation of two-time correlations in photon emissions from a single atom, *Phys. Rev. A.* 18, 2217 (1978).
[20] R. L. Pfleegor and L. Mandel, Interference of independent photon beams, *Phys. Rev.* 159, 1084 (1967).

Chapter 2

Enhanced Vector Potential Quantization and Electromagnetic Vacuum*

Abstract

We investigate the quantization of the vector potential of the electromagnetic field enhanced to a single photon state, starting from the fundamental equations linking the energy density of the classical electromagnetic theory to the quantum electrodynamics expressions. The resulting wave-particle formalism ensures a coherent transition between the classical electromagnetic wave theory and the QED representation. Following this procedure, a quantization constant of the photon vector potential amplitude is naturally defined, entailing a new description of the electromagnetic quantum vacuum without energy singularity. Applying the established vacuum formalism, the calculated spontaneous emission rate and Lamb shift for the atomic hydrogen nS states are in agreement with the experimental evidence.

1. Theoretical background of the QED vacuum state

It is well known [1–5] that the electromagnetic field Hamiltonian in QED is written as

$$H = \sum_{k,\lambda} \left(\hat{n}_{k\lambda} + \frac{1}{2} \right) \hbar\omega_k \tag{1}$$

*This chapter is reproduced from C. Meis, 2014. *Physics Research International*, Vol 2014, 187432, 1–5.

where $\hbar$ is Planck's reduced constant $(h/2\pi)$ and $\hat{n}_{k\lambda} = a^+_{k\lambda} a_{k\lambda}$ is the Hermitian photon number operator, where $a^+_{k\lambda}$ and $a_{k\lambda}$ are the creation and annihilation non-Hermitian operators of a k-mode and λ-polarization photon, respectively, with angular frequency ω_k. In the absence of photons, (1) leads to the well-known QED zero-point energy singularity corresponding physically to infinite vacuum energy [5–7].

It has been demonstrated that even when considering only the QED quantum vacuum and applying quite "reasonable" frequency cut-offs, the obtained energy density is roughly 120 orders of magnitude greater than the one actually observed [8, 9] in the universe. This situation has been called the "quantum vacuum catastrophe," entailing the necessity of new theoretical developments. Indeed, in the last two decades, various studies, developing different models [10–19], have been carried out in order to interpret the recent astrophysical observations, which are in contradiction with the quantum field theory [5–8]. Those studies have revealed the necessity of developing a different approach for the electromagnetic vacuum representation.

In this paper, we enhance the quantization procedure of the electromagnetic field vector potential [20] down to a single photon level in order to better understand the nature of the photon and that of the vacuum state. The resulting quantum vacuum field fluctuations correspond to a very low energy density medium that can be compatible with the astrophysical observations. It is also expressed as a function of the single photon creation and annihilation operators and is further applied in this paper for the calculation of the spontaneous emission transition rates as well as the Lamb shift of the atomic hydrogen ηS states, showing the consistency with the experimental evidence.

2. Quantized vector potential of a single photon

We will not consider here the zero-point energy singularity $\sum_{k,\lambda} \hbar\omega_k/2$ obtained from (1) in the absence of photons since this term drops when using the "normal ordering" for the photon creation and annihilation operators in the quantization process [5,6,21]. Furthermore, it has been demonstrated [22] that the Casimir effect can very well be calculated without referring at all to the zero-point energy. Hence, we start from the fundamental relations defining the transition between the classical electromagnetic wave vector potential amplitude and the corresponding quantized field operators [1–7] used in QED:

$$A_{k\lambda} \to \sqrt{\frac{\hbar}{2\varepsilon_0\omega_k V}}a_{k\lambda}; \quad A^*_{k\lambda} \to \sqrt{\frac{\hbar}{2\varepsilon_0\omega_k V}}a^+_{k\lambda} \tag{2}$$

where ϵ_0 is the vacuum electric permittivity and V is the quantization volume. It is very important to recall that the above link equations result from the energy density equivalence as expressed by the classical electromagnetic wave theory, based on Maxwell's equations, and the representation of photons in QED.

In fact, the mean energy density over a period $2\pi/\omega_k$ of a classical monochromatic electromagnetic plane wave with angular frequency ω_k has been "imposed" to be equal to the energy density of n_k photons in a given volume V according to the relation [1]

$$\begin{aligned} &\left\langle \frac{1}{2}\left(\varepsilon_0\left|\vec{E}_k(\vec{r},t)\right|^2 + \frac{1}{\mu_0}\left|\vec{B}_k(\vec{r},t)\right|^2\right)\right\rangle_{\text{period}} \\ &= \left\langle 4\varepsilon_0\omega_k^2 A_{0k}^2(\omega_k)\sin^2(\vec{k}\cdot\vec{r} - \omega_k t + \phi)\right\rangle_{\text{period}} \\ &= 2\varepsilon_0\omega_k^2 A_{0k}^2(\omega_k) \equiv \frac{n_k\hbar\omega_k}{V} \end{aligned} \tag{3}$$

where $\vec{E}_k(\vec{r},t)$ and $\vec{B}_k(\vec{r},t)$ are the electric and magnetic fields, respectively, of the k-mode plane wave, with ε_0 and μ_0 the electric permittivity and magnetic permeability of the vacuum, respectively, satisfying the relation $\varepsilon_0\mu_0 c^2 = 1$, where c is the speed of light in vacuum. Also, $A_{0k}(\omega_k)$ is the vector potential amplitude of the k-mode plane electromagnetic wave. Therein, it is worth noting that the volume V is an external parameter and the link relations (2) used in QED are obtained from (3) for $n_k = 1$, i.e., for a single photon state. The fact that the photon vector potential depends on an external arbitrary parameter V is rather puzzling.

Considering that the vector potential generally consists of a superposition of k modes and λ polarizations, it is written with the well-known expression as

$$\vec{A}(\vec{r},t) = \sum_k \sqrt{\frac{\hbar}{2\varepsilon_0 \omega_k V}} \sum_{\lambda=1}^{2} \Big[\alpha_{k\lambda}\hat{\varepsilon}_{k\lambda} e^{i(\vec{k}\cdot\vec{r}-\omega_k t+\theta)} + \alpha^{+}_{k\lambda}\hat{\varepsilon}^{*}_{k\lambda} e^{-i(\vec{k}\cdot\vec{r}-\omega_k t+\theta)}\Big] \tag{4}$$

where the summation runs over all the modes k and the two polarizations, circular left and right, $\hat{\varepsilon}_{k\lambda}$ is the complex unit vector of the polarization and θ is the phase parameter.

Recent experiments [23, 24] demonstrated the individuality and indivisibility of photons. However, it is obvious that the link equations used above for the vector potential amplitude do not result in a well-defined description of individual photons, but they are rather valid for a large number of photons in a big volume V. This is precisely the reason why in QED, based on the density of states theory, the discrete summation over the modes is often replaced by a continuous summation over the frequencies:

$$\sum_k \to \frac{V}{8\pi^3} \int 4\pi k^2 dk = \frac{V}{2\pi^2 c^3} \int \omega^2 d\omega \tag{5}$$

eliminating by this way the parameter V in all calculations involving the square of the vector potential. The expression (5) is valid whatever the shape of the volume V is, provided that the dimensions of the last one are much bigger than the electromagnetic field wavelengths [2, 5, 25]. Let us now have a close look at the vector potential amplitude of a single photon. Therein, we recall that a dimensional analysis of the general expression of the vector potential amplitude of an electromagnetic field shows that it should be proportional to the angular frequency ω_k. Note that this is not a hypothesis but a mathematical deduction from Maxwell's equations.

Consequently, for the vector potential of a k-mode photon, we should have [20]

$$\alpha_{0k}(\omega_k) \propto (\text{Constant} \times \omega_k) \to \xi\omega_k \tag{6}$$

So, the expression of the vector potential assumed to be that of a single photon may now be written in a general plane wave expression

$$\vec{\alpha}_k(\omega_k,\vec{r},t) = \xi\omega_k \Big[\hat{\varepsilon}_{k\lambda} e^{i(\vec{k}\cdot\vec{r}-\omega_k t+\theta)} + \hat{\varepsilon}^{*}_{k\lambda} e^{-i(\vec{k}\cdot\vec{r}-\omega_k t+\theta)}\Big] \tag{7}$$

whose propagation takes place within a period T over a wavelength λ_k and repeated successively along the propagation axis. It has been demonstrated [20] that $\vec{\alpha}_k(\omega_k, \vec{r}, t)$ satisfies the wave propagation equation

$$\frac{\partial^2}{\partial t^2}\vec{\alpha}_k(\omega_k, \vec{r}, t) - c^2\vec{\nabla}^2\vec{\alpha}_k(\omega_k, \vec{r}, t) = 0 \tag{8}$$

as well as a linear time differential equation

$$i\xi\frac{\partial}{\partial t}\vec{\alpha}_k(\omega_k, \vec{r}, t) = \tilde{\alpha}_{0k}\left(\omega_k\right)\vec{\alpha}_k(\omega_k, \vec{r}, t) \tag{9}$$

where the vector potential amplitude operator is $\tilde{\alpha}_{0k}(\omega_k) = -i\xi c\vec{\nabla}_k$.

The mathematical correspondence between the pairs $[E_k, \hbar]$ and $[\alpha_{0k}, \xi]$ is obvious.

Following Eq. (7), the photon vector potential functions can be expressed with both wave and particle formalism:

$$\begin{aligned}\vec{\alpha}_k(\omega_k, \vec{r}, t) &= \xi\omega_k\left[\hat{\varepsilon}_{k\lambda}e^{i(\vec{k}\cdot\vec{r}-\omega_k t+\theta)} + \hat{\varepsilon}^*_{k\lambda}e^{-i(\vec{k}\cdot\vec{r}-\omega_k t+\theta)}\right]\\ &= \omega_k\vec{\Xi}_{k\lambda}(\omega_k, \vec{r}, t) \end{aligned} \tag{10a}$$

$$\begin{aligned}\tilde{\alpha}_{k\lambda} &= \xi\omega_k\left[a_{k\lambda}\hat{\varepsilon}_{k\lambda}e^{i(\vec{k}\cdot\vec{r}-\omega_k t+\theta)} + a^+_{k\lambda}\hat{\varepsilon}^*_{k\lambda}e^{-i(\vec{k}\cdot\vec{r}-\omega_k t+\theta)}\right]\\ &= \omega_k\tilde{\Xi}_{k\lambda}(\omega_k, \vec{r}, t) \end{aligned} \tag{10b}$$

where, for the quantum mechanical formalism, we have used the creation and annihilation operators for a k-mode and λ-polarization photon.

The mathematical representations of the wave and particle expressions of the vector potential amplitude operator is written as [20]

$$\text{wave:}\ \tilde{\alpha}_{0k} = -i\xi c\vec{\nabla}_k;\quad \tilde{\alpha}^*_{0k} = i\xi c\vec{\nabla}_k \tag{11a}$$

$$\text{particle:}\ \tilde{\alpha}_{0k\lambda} = \xi\omega_k a_{k\lambda};\quad \tilde{\alpha}^+_{0k\lambda} = \xi\omega_k a^+_{k\lambda} \tag{11b}$$

A calculation of the vector potential amplitude quantization constant ξ has been carried out previously [26]:

$$\xi = \pm\frac{1}{(2\pi)^{3/2}}\sqrt{\frac{\hbar}{8\alpha\varepsilon_0 c^3}} = \frac{\hbar}{4\pi ec} = \pm 1.747\,10^{-25}\,\mathrm{V\,m^{-1}\,s^2} \tag{12}$$

where α is the fine structure constant and e is the electron–positron elementary charge.

Appropriate experiments could attribute a more precise value to this constant.

3. The electromagnetic vacuum field

According to the relations (10), for $\omega_k = 0$, the vector potential as well as the energy of a k-mode photon vanish. However, the principal function of the vector potential $\Xi_{k\lambda}(\omega_k = 0, \vec{r}, t)$ does not vanish but is reduced to the field

$$\tilde{\Xi}_{0\lambda}(\omega_k = 0, \vec{r}, t) = \xi\left(\hat{\varepsilon}_{k\lambda} a_{k\lambda} e^{i\theta} + \hat{\varepsilon}^*_{k\lambda} a^+_{k\lambda} e^{-i\theta)}\right) \tag{13}$$

Thus, $\Xi_{0\lambda}(\omega_k = 0, \vec{r}, t)$ is a real field having the physical dimensions Volt $\mathrm{m}^{-1}\,\mathrm{s}^2$, corresponding to a vector potential per angular frequency. At zero energy and zero vector potential, $\Xi_{0\lambda}(\omega_k = 0, \vec{r}, t)$ represents the electromagnetic vacuum field. Obviously, it's a dynamic entity capable of inducing electronic transitions in matter since it is described by a function of the photon creation and annihilation non-Hermitian operators. In the following, we apply the established electromagnetic vacuum expression for the calculation of the spontaneous emission rate and the Lamb shift of the atomic hydrogen nS states.

3.1. *Spontaneous emission*

In a rigorous calculation within the standard QED theory, the interaction Hamiltonian between an atomic electron with dipole moment $\vec{d} = e\vec{r}$ and the vacuum is written as $H_{\text{int}} = -\vec{d} \cdot \left(\frac{i}{\hbar}\right)\left[\vec{A}_{k\lambda}(\vec{r}, t), \sum_{k,\lambda} \frac{1}{2}\hbar\omega_k\right]$, which vanishes for the simple physical reason that the zero-point energy is a constant and is not composed of a function of the photon creation and annihilation operators.

Consequently, an interaction Hamiltonian for the spontaneous emission cannot be defined in the standard QED theory. In fact, it is well known that in QED, the spontaneous emission rate as well as the Lamb shift are calculated based on the commutation property $[a_{k\lambda}, a^+_{k'\lambda'}] = \delta_{kk'}\delta_{\lambda\lambda'}$ of the photon creation and annihilation operators without invoking at all the zero-point energy term $\sum_{k,\lambda} \hbar\omega_k/2$.

Conversely, for the description of the electron–vacuum interaction, a "vacuum action" operator corresponding to a Hamiltonian interaction per angular frequency between the vacuum field $\Xi_{0\lambda}(\omega_k = 0, \vec{r}, t)$ and an atomic electron of mass m_e and charge e can now be defined:

$$H_k = -i\hbar \frac{e}{m_e} \tilde{\Xi}_{0\lambda} \cdot \vec{\nabla}_k \tag{14}$$

Thus, the amplitude of the transition probability per angular frequency between an initial state $|\Psi_i, 0\rangle$, with total energy $E_i = \hbar\omega_i$, corresponding to an atom in the state $|\Psi_i\rangle$ with energy E_i in vacuum $|\ldots, 0_{k\lambda}, \ldots, 0_{k'\lambda'}, \ldots\rangle$, and a final state $|\Psi_f, n_{k\lambda}\rangle$ with total energy $E_{f,nk\lambda} = \hbar\omega_f + n_{k\lambda}\hbar\omega_k$, representing the atom in the state $|\Psi_f\rangle$ with energy $E_f = \hbar\omega_f$ in the presence of $n_{k\lambda}$ photons, can be expressed in first-order time-dependent perturbation theory using the well-known formalism [1, 3, 5]

$$\begin{aligned} c^{(1)}_{\omega_k} = -\frac{e}{m_e} \int_0^t & [\langle \Psi_f, n_{k\lambda} | \xi \hat{\varepsilon}_{k\lambda} \cdot \vec{\nabla}_k a_{k\lambda} | \Psi_i, 0\rangle e^{i(\omega_i - \omega_f - n_{k\lambda}\omega_k)t'} \\ & + \langle \Psi_f, n_{k\lambda} | \xi \hat{\varepsilon}^*_{k\lambda} \cdot \vec{\nabla}_k a^+_{k\lambda} | \Psi_i, 0\rangle e^{-i(\omega_i - \omega_f - n_{k\lambda}\omega_k)t'}] dt' \end{aligned} \tag{15}$$

As $a_{k\lambda}|\Psi_i, 0\rangle = 0$, using Heisenberg's equation of motion [5, 21], the scalar product of the last equation involving the creation operator is written as

$$\langle \Psi_f, n_{k\lambda} | \xi \hat{\varepsilon}^*_{k\lambda} \cdot \vec{\nabla}_k a^+_{k\lambda} | \Psi_i, 0\rangle = -\xi \delta_{1,n_{k\lambda}} m_e \omega_{if} \hat{\varepsilon}^*_{k\lambda} \cdot \frac{\vec{r}_{if}}{h} \tag{16}$$

with $\vec{r}_{if} = \langle \Psi_f | \vec{r} | \Psi_i\rangle$, $\omega_{if} = \omega_i - \omega_f$ and $\delta_{1,n_{k\lambda}}$, the Kronecker's delta symbol.

Using now $\xi = \frac{\hbar}{4\pi ec}$ from Eq. (12), we get the spontaneous emission rate within a solid angle $d\Omega$:

$$W_{if} = \frac{1}{3\hbar c^3} \frac{e^2}{4\pi\varepsilon_0} \omega^3_{if} |\vec{r}_{if}|^2 d\Omega \tag{17}$$

This result is in full agreement with the previous QED calculations [1, 3, 21] and shows that the spontaneous emission is mainly due to the creation operator, which in this formalism is involved in the electromagnetic vacuum expression and the interaction Hamiltonian. Consequently, $\Xi_{0\lambda}(\omega_k = 0, \vec{r}, t)$ is the "skeleton" of every photon. According to the relations (10), a photon appears to be a vacuum oscillation generated when the last one interacts with matter, conferring an angular frequency ω_k to the vacuum state $\Xi_{0\lambda}(\omega_k = 0, \vec{r}, t)$.

3.2. *Lamb shift of the atomic hydrogen nS levels*

Various calculations of the energy shift of the nS levels of the hydrogen atom have been carried out, each one by considering the vacuum effect on the atomic orbitals through different aspects, mass renormalization [27], vacuum electric field perturbation [28], Feynman's gas interpretation [29], Stark shift [5], and so forth. Here, we advance a quite simple and complementary approach based on the $\Xi_{0\lambda}(\omega_k = 0, \vec{r}, t)$ electromagnetic vacuum field representation. In the first approximation, the energy level displacements of the electron-bound states can be seen as a topological effect of the vacuum radiation pressure exerted upon the electronic orbitals. In fact, the motion of a bounded state electron with charge e, whose energy is E_{nlj} in the vacuum field $\Xi_{0\lambda}(\omega_k = 0, \vec{r}, t)$ can be characterized by Bohr's angular frequency

$$\omega_{nlj} = \frac{E_{nlj}}{\hbar} = \frac{R_\infty}{n^2\hbar} \tag{18}$$

where n, l, and j are the quantum numbers of the corresponding orbital and where $R_\infty = 13.606$ eV. The oscillation motion of the atomic electron in the electromagnetic vacuum field yields the rise in the electron frame of a vector potential amplitude

$$\alpha_{0(nlj)} = \xi\omega_{nlj} \tag{19}$$

Corresponding to the vector potential of vacuum photons with energy,

$$4\pi ec\,\xi\,\omega_{nlj} = \hbar\omega_{nlj} \tag{20}$$

Hence, due to its periodic motion in the field $\Xi_{0\lambda}(\omega_k = 0, \vec{r}, t)$, an electron experiences in its frame vacuum photons whose pressure per unit surface is written as [25]

$$dP_{vac} = \sum_\lambda \varepsilon_0 |E_0^{(R,L)}|^2 d\Omega \tag{21}$$

where $E_0^{(R,L)}$ is the electric field amplitude of the right (R) and left (L) hand polarized vacuum photons seen by the electron at the same oscillation frequency.

The electric field amplitude can be readily calculated [1,25] and written as

$$\left|E_0^{(R,L)}\right|^2 = \omega_{nlj}^2\left(\left|\alpha_{0x}^{(R)}\right|^2 + \left|\alpha_{0y}^{(R)}\right|^2\right) = \omega_{nlj}^2\left(\left|\alpha_{0x}^{(L)}\right|^2 + \left|\alpha_{0y}^{(L)}\right|^2\right)$$
$$= \omega_{nlj}^2\left(2\xi^2\omega_{nlj}^2\right) \quad (22)$$

The summation over the two circular polarizations and over the whole solid angle 4π gives the total vacuum radiation pressure upon the nlj atomic orbital:

$$P_{vac} = 4\pi\varepsilon_0(4\xi^2\omega_{nlj}^4) = \frac{\hbar\omega_{nlj}^4}{4\alpha\pi^2c^3} \quad (23)$$

Hence, if the electronic cloud subsists in the effective volume $V_{nlj}(\text{eff})$ of the nlj orbital, the energy shift due to the vacuum radiation pressure is written as

$$\delta E = P_{vac}V_{nlj}(\mathit{eff}) = \frac{\hbar\omega_{nlj}^4}{4\alpha\pi^2c^3}V_{nlj}(\mathit{eff}) \quad (24)$$

A delicate operation consists of defining $V_{nlj}(\text{eff})$. The nS electronic orbitals, for example, are characterized by a probability distribution density decreasing smoothly with an exponential expression. Furthermore, for the excited electronic nS states, the radial wave functions contain negative values, and the electron probability density distribution involves a significant space area with zero values. Hence, in the case of atomic hydrogen for the spherically symmetrical electronic orbitals nS, in order not to take into account the space regions where the probability density is zero, the effective volume may be written in the first approximation as

$$V_{nS}(\mathit{eff}) \approx \frac{1}{n}\frac{4}{3}\pi|\langle\Psi_{nS}|\vec{r}|\Psi_{nS}\rangle|^3 \quad (25)$$

Putting $\omega_{nS} = R_\infty/n^2\hbar$ and considering the cube of the mean value of the distance in the atomic orbitals nS,

$$|\langle r\rangle_{n,l=0}|^3 = \left|\int d^3r\Psi_{n,l=0}r\Psi_{n,l=0}^*\right|^3 = a_0^3\left(\frac{3n^2}{2}\right)^3 \quad (26)$$

with $a_0 = 0.53\,10^{-10}\,\text{m}$ the first Bohr radius, the frequency of the energy shift is written in the first approximation as

$$\nu_{nS} \approx \frac{1}{n^3}\frac{9R_\infty^4 a_0^3}{16\alpha\pi^2 c^3\hbar^4} \tag{27}$$

We obtain the following values for the Lamb shift frequencies for the atomic hydrogen nS levels:

$$\begin{aligned}\nu_{1S} &\sim 7.96\,\text{GHz}\ (8.2)\\ \nu_{2S} &\sim 1000\,\text{MHz}\ (1040)\\ \nu_{3S} &\sim 297\,\text{MHz}\ (303)\\ \nu_{4S} &\sim 129\,\text{MHz}\ (131)\end{aligned} \tag{28}$$

with the experimental values given in parentheses.

The agreement with the experiment is quite satisfactory for such a simple calculation.

3.3. *Astrophysical observations and the quantum vacuum energy density*

It has been pointed out that recent astronomical observations revealed the vacuum energy density in the universe to be roughly 120 orders of magnitude [5, 8] lower than that predicted by quantum field theory. In fact, the studies on dark energy and the arising constraints have stimulated the creation of new models to remedy the theoretical difficulties in cosmology. The modeling of the dark energy as a fluid has led to the classical phantom description [10], which was revealed to be unstable from both quantum mechanical and gravitational points of view. Some proposals, among many others, based on phantom cosmology have introduced more consistent dark energy models by introducing quantum effects in general relativity [11], nonlinear gravity-matter interaction [12], revised gravity [13], and effective phantom phase and holography [14]. On the other hand, various studies [15–17] on the contribution of the quantum vacuum state to the dark energy have been carried out based on both quantum chromodynamics (QCD) and quantum electrodynamics (QED) descriptions. However, it has been pointed out by many authors that our understanding of the QCD vacuum state still remains very elusive. Conversely, the QED vacuum offers a more comprehensive framework, and there is a general scientific effort to analyze the contribution of the electromagnetic field's lower energy level to the dark energy. For a finite universe, which we may consider behaving as a cosmic cavity, the electromagnetic vacuum field $\Xi_{0\lambda}(\omega_k = 0, \vec{r}, t)$ oscillation frequency gets very small but not zero. Thus, we may assume that the universe is filled with a superposition of photons corresponding to a very low energy density. Under these conditions, an extremely high number of cosmic electromagnetic waves, issued from the fluctuations of the vacuum states, could correspond to the measured background vacuum energy density, which is close to $10^{-9}\,\mathrm{J/m^3}$. Hence, the energy density corresponding to the fluctuations of the vacuum field $\Xi_{0\lambda}(\omega_k = 0, \vec{r}, t)$ is worthy to be calculated and compared to the astrophysical observations.

4. Conclusion

The difficulties raised in QED and in cosmology by the infinite energy of the zero-point energy can be lifted by considering the quantization of the electromagnetic field vector potential amplitude to a single photon $\alpha_{0k} = \xi\omega_k$. This fact is not a model based on any assumptions or postulates but results directly from Maxwell's equations. The established theoretical developments are in agreement with the experimental evidence according to which the photon is revealed to be an integral and indivisible particle with simultaneous wave-particle properties. We have thus obtained both wave and particle representations of the vector potential whose amplitude is proportional to a constant ξ. The electromagnetic vacuum field appears naturally and is expressed through the photon creation and annihilation operators, entailing that it is a dynamic entity capable of inducing electronic transitions in atoms and molecules. We have shown that the spontaneous emission rate and the Lamb shift of the atomic hydrogen nS orbitals can equally be calculated using the electromagnetic vacuum expression $\tilde{\Xi}_{0\lambda}(\omega_k = 0, \vec{r}, t) = \xi(\hat{\varepsilon}_{k\lambda} a_{k\lambda} e^{i\theta} + \hat{\varepsilon}^*_{k\lambda} a^+_{k\lambda} e^{-i\theta)})$.

Finally, the energy density of the electromagnetic vacuum fluctuations might be compatible with recent vacuum measurements and is worthy to be calculated. This opens perspectives for new experiments in order to obtain a more precise value for the constant ξ and also for further astrophysical observations with the aim of getting an evaluation of the density of very low energy cosmic photons contributing to the background energy density. However, detailed calculations on the electromagnetic vacuum fluctuations and the corresponding mean energy density created in space are presented in Chapter 9.

References

[1] B. H. Bransden and C. J. Joachain, *Physics of Atoms and Molecules* (Longman, New York, NY, USA, 1983).

[2] L. H. Ryder, *Quantum Field Theory* (Cambridge University Press, Cambridge, UK, 1987).

[3] H. Haken, *Light* (North Holland Publishing, Amsterdam, The Netherlands, 1981).

[4] A. I. Akhiezer and B. V. Berestetskii, *Quantum Electrodynamics* (Interscience Publishers, New York, NY, USA, 1965).

[5] P. W. Milonni, *The Quantum Vacuum* (Academic Press, Boston, MA, USA, 1994).

[6] C. P. Enz, Is the zero-point energy real? in *Physical Reality and Mathematical Description*, C. P. Enz and J. Mehra, Eds. (Reidel, Dordrecht, The Netherlands, 1974).

[7] R. P. Feynman, *The Strange Theory of Light and Matter* (Princeton University Press, Princeton, NJ, USA, 1988).

[8] S. Weinberg, The cosmological constant problem, *Rev. Mod. Phys.* 61(1), 1–23 (1989).

[9] T. Hey and P. Walters, *The New Quantum Universe* (Cambridge University Press, Cambridge, UK, 2003).

[10] S. M. Carroll, M. Hoffman, and M. Trodden, Can the dark energy equation-of-state parameter w be less than −1? *Phys. Rev. D* 68(2), 023509 (2003).

[11] V. Fanaoni, Cumulative author index B701–B703, *Nucl. Phys. B* 703(3), 557–558 (2004).

[12] S. Nojiri and S. D. Odintsov, Gravity assisted dark energy dominance and cosmic acceleration, *Phys. Lett. B* 599(3-4), 137–142 (2004).

[13] M. C. B. Abdalla, S. Nojiri, and S. D. Odintsov, Consistent modified gravity: dark energy, acceleration and the absence of cosmic doomsday, *Class. Quantum Gravity* 22(5), L35–L42 (2005).

[14] E. Elizalde, S. Nojiri, S. D. Odintsov, and P. Wang, Dark energy: Vacuum fluctuations, the effective phantom phase, and holography, *Phys. Rev. D* 71(10), 103504 (2005).
[15] J. A. Frieman, M. S. Turner, and D. Huterer, Dark energy and the accelerating universe, *Annu. Rev. Astron. Astrophys.* 46, 385–432 (2008).
[16] A. Silvestri and M. Trodden, Approaches to understanding cosmic acceleration, *Rep. Prog. Phys.* 72(9), 096901 (2009).
[17] L. Labun and J. Rafelski, Vacuum structure and dark energy, *Int. J. Mod. Phys. D* 19(14), 2299–2304 (2010).
[18] P. J. E. Peebles and B. Ratra, The cosmological constant and dark energy, *Rev. Mod. Phys.* 75(2), 559–606 (2003).
[19] T. Padmanabhan, Cosmological constant—the weight of the vacuum, *Phys. Rep.* 380(5-6), 235–320 (2003).
[20] C. Meis, Photon wave-particle duality and virtual electromagnetic waves, *Found. Phys.* 27(6), 865–873 (1997).
[21] M. Weissbluth, *Photon-Atom Interactions* (Academic Press, Boston, 1988).
[22] P. W. Milonni, Casimir forces without the vacuum radiation field, *Phys. Rev. A.* 25(3), 1315–1327 (1982).
[23] P. Grangier, G. Roger, and A. Aspect, Experimental evidence for a photon anticorrelation effect on a beamsplitter: A new light on single-photon interferences, *Europhys. Lett.* 1(4), 173 (1986).
[24] J. C. Garrison and R. Y. Chiao, *Quantum Optics* (Oxford University Press, Oxford, UK, 2008).
[25] F. H. Read, *Electromagnetic Radiation* (John Wiley & Sons, New York, NY, USA, 1980).
[26] C. Meis, Electric potential of the quantum vacuum, *Phys. Essays.* 22(1), 17–18 (2009).
[27] H. A. Bethe, The electromagnetic shift of energy levels, *Phys. Rev.* 72(4), 339–341 (1947).
[28] T. A. Welton, Some observable effects of the quantum-mechanical fluctuations of the electromagnetic field, *Phys. Rev.* 74(9), 1157–1167 (1948).
[29] R. P. Feynman, *The Quantum Theory of Field* (Wiley Interscience, New York, 1961).

Chapter 3

Vector Potential Quantization and Photon Intrinsic Electromagnetic Properties: Toward Non-destructive Photon Detection*

Abstract

We employ here the enhancement of the vector potential amplitude quantization at a single photon state. The analysis of the general solution of the vector potential, obtained by resolving Maxwell's equations, implies that the amplitude is proportional to the angular frequency. The photon vector potential function $\vec{\alpha}_{k\lambda}(\vec{r}, t)$ can be written in the plane wave representation, satisfying the classical wave propagation equation, Schrödinger's equation for the energy with the relativistic massless field Hamiltonian and a linear time-dependent equation for the vector potential amplitude operator. Thus, the vector potential $\vec{\alpha}_{k\lambda}(\vec{r}, t)$ with the quantized amplitude may play the role of a real wave function for the photon in a non-local representation that can be suitably normalized. We then deduce that the amplitudes of the electric and magnetic fields, respectively, of a single free photon are proportional to the square of the angular frequency. This might open perspectives for the development of non-destructive photon detection methods based on the influence of the electric and/or magnetic fields of photons on the energy levels of atoms and molecules.

*This chapter is reproduced from C. Meis and P. R. Dahoo, 2017. "Vector potential quantization and the photon intrinsic electromagnetic properties: Towards nondestructive photon detection", *International Journal of Quantum Information*, 15(8), 1740003.

1. Introduction

In recent years, the experimental evidence of the entangled states produced with single photon sources has demonstrated the permanent violation of Bell's inequality, entailing the necessity of developing a non-local representation for the photon, through a real wave function [1–8]. At the same time, the existence of hidden variables within a local representation is rejected, while new variables in a non-local representation are not explicitly excluded [3, 6, 9]. Hence, we investigate in this paper the possibility of expressing the single free photon intrinsic electromagnetic properties within a non-local representation [10–13] through the vector potential with quantized amplitude. In fact, a detailed unit analysis of the general solution of Maxwell's equations for the vector potential [14–16] shows that the amplitude is proportional to the angular frequency. Thus, for a k-mode photon with angular frequency ω_k, we may write the vector potential amplitude as $\alpha_{0k} = \xi\omega_k$, where ξ is a constant.

In the first approximation, the value of the constant ξ has been evaluated [14, 16–18] by normalizing the mean energy of a plane electromagnetic wave over a wavelength to a single photon energy, obtaining

$$|\xi| = \left|\frac{1}{(2\pi)^{3/2}}\sqrt{\frac{\hbar}{8\alpha\,\varepsilon_0 c^3}}\right| = \left|\frac{\hbar}{4\pi\,e\,c}\right| = 1.747\ \ 10^{-25}\ \mathrm{Vm^{-1}s^2} \qquad (1)$$

with $\alpha = 1/137$ the fine structure constant, ε_0 the vacuum electric permittivity, c the speed of light in vacuum and e the electron-positron charge.

The fundamental physical properties characterizing the wave–particle nature of a single k-mode photon [14–18]: {energy E_k and momentum $\vec{p}_k$} (particle), {vector potential amplitude α_{0k}, wave vector $\vec{k}$ and dispersion relation} (wave), are all related directly to the angular frequency ω_k:

$$\frac{E_k}{\hbar} = \frac{|\vec{p}_k|}{\hbar/c} = \frac{\alpha_{0k}}{\xi} = |\vec{k}|c = \omega_k \qquad (2)$$

The last relation expresses the wave–particle properties of a single free photon state, introducing the electromagnetic nature through the quantized vector potential amplitude.

2. Photon vector potential within non-local representation and intrinsic electromagnetic fields

The relation (2) entails that the quantized vector potential amplitude is an intrinsic property of a single photon. Hence, the vector potential for a k-mode and λ-polarization photon can now be written in the plane wave representation

$$\vec{\alpha}_{k\lambda}(\vec{r},t) = \xi\omega_k \left(\hat{\varepsilon}_{k\lambda}e^{i(\vec{k}\cdot\vec{r}-\omega_k t+\theta)} + \hat{\varepsilon}^*_{k\lambda}e^{-i(\vec{k}\cdot\vec{r}-\omega_k t+\theta)}\right) \tag{3}$$

where λ takes two values corresponding to the left (L) and right (R) circular polarizations expressed by the unit vector $\varepsilon_{k\lambda}$.

The general expression for the vector potential of an electromagnetic wave considered as a superposition of different modes is written as

$$\vec{A}(\vec{r},t) = \sum_{k,\lambda} \xi\omega_k \left[\hat{\varepsilon}_{k\lambda}e^{i(\omega_k t-\vec{k}\cdot\vec{r}+\theta)} + \hat{\varepsilon}^*_{k\lambda}e^{-i(\omega_k t-\vec{k}\cdot\vec{r}+\theta)}\right] \tag{4}$$

It can be easily shown [14,16,18,19] that the vector potential function (3) satisfies the classical wave propagation equation in vacuum

$$\vec{\nabla}^2\vec{\alpha}_{k\lambda}(\vec{r},t) - \frac{1}{c^2}\frac{\partial^2}{\partial t^2}\vec{\alpha}_{\kappa\lambda}(\vec{r},t) = 0 \tag{5}$$

as well as Schrödinger's equation with the well-known relativistic massless particle Hamiltonian $\tilde{H} = -i\hbar c\vec{\nabla}$

$$i\,\hbar\frac{\partial}{\partial t}\vec{\alpha}_{k\lambda}(\vec{r},t) = \tilde{H}\,\vec{\alpha}_{k\lambda}(\vec{r},t) \tag{6}$$

having the eigenvalue of a single photon energy.

The vector potential function (3) also satisfies a linear time-dependent equation for the vector potential amplitude operator [14–16]

$$\tilde{\alpha}_{0k} = -i\xi\,c\vec{\nabla}_k \tag{7}$$

which is written as

$$i\,\xi\frac{\partial}{\partial t}\vec{\alpha}_{k\lambda}(\vec{r},t) = \tilde{\alpha}_{0k}\,\vec{\alpha}_{k\lambda}(\vec{r},t) \tag{8}$$

where the gradient subscript k merely denotes that it acts upon the mode k.

The symmetrical equations for the energy and the vector potential, (6) and (8), respectively, can be combined to obtain the vector *potential–energy* (*wave–particle*) equation for the photon

$$i\begin{pmatrix}\xi\\ \hbar\end{pmatrix}\frac{\partial}{\partial t}\vec{\alpha}_{k\lambda}(\vec{r},t)=\begin{pmatrix}\tilde{\alpha}_{0k}\\ \tilde{H}\end{pmatrix}\vec{\alpha}_{k\lambda}(\vec{r},t) \tag{9}$$

having the eigenvalues $\xi\omega_k$ and $\hbar\omega_k$, respectively.

The photon as an integral entity is not a point but extends naturally over a wavelength according to the experimental evidence [3,20,21] in agreement with the density of states theory [3,18,22–24]. The wave function of the quantized vector potential (3) ensures that the photon as a whole propagates along the propagation axis with circular (L or R) polarization rotating at the angular frequency ω_k.

For a cavity-free k-mode photon, we can deduce the amplitude of the intrinsic electric field $|\vec{\varepsilon}_{k\lambda}|$, which appears to be proportional to the square of the angular frequency

$$|\vec{\varepsilon}_{k\lambda}|=\left|-\frac{\partial}{\partial t}\vec{\alpha}_{k\lambda}(\vec{r},t)\right|\propto\xi\omega_k^2 \tag{10}$$

According to the plane wave representation, the corresponding magnetic field $|\vec{\beta}_{k\lambda}|$ of the free photon is equally proportional to the square of the angular frequency

$$|\vec{\beta}_{k\lambda}|\propto\sqrt{\varepsilon_0\mu_0}\,\xi\omega_k^2 \tag{11}$$

where μ_0 is the vacuum magnetic permeability.

It is of fundamental interest to experimentally investigate the latest results. High-resolution laser spectroscopy might be employed for the measurement of the Stark shifts induced by the electric field of an energetic photon (e.g., in the UV range) on the atomic or molecular energy levels. Such an experimental investigation could also confirm the dependence of the single photon electric field on the square of the angular frequency. In fact, the energy shift due to the linear Stark effect δE_{Stark} is directly proportional to the local electric field strength, which for the photon depends on the angular frequency, $\delta E_{\text{Stark}}\propto|\vec{\varepsilon}_{k\lambda}(\omega_k)|\propto\omega_k^p$. The ratio of the measured Stark shifts

$\delta E^{(1)}_{\text{Stark}}$ and $\delta E^{(2)}_{\text{Stark}}$, induced independently by two photons with quite different energies $\hbar\omega_{k1}$ and $\hbar\omega_{k2}$, could permit us to define directly the dependence of the photon electric field on the angular frequency according to the expression $\frac{\delta E^{(1)}_{\text{Stark}}}{\delta E^{(2)}_{\text{Stark}}} \propto \left(\frac{\omega_{k1}}{\omega_{k2}}\right)^p$, i.e., if $p = 2$, as deduced by the relation (10), then for $\omega_{k1} = 2\omega_{k2}$, the ratio of the measured Stark shifts should be roughly $\delta E^{(1)}_{\text{Stark}} \approx 4\delta E^{(2)}_{\text{Stark}}$. Such an experimental confirmation might also open perspectives for the development of non-destructive photon detection methods through the measurement of the influence of the photon electric or magnetic fields upon the atomic or molecular energy levels.

3. Conclusion

Within a non-local photon representation, as required by entangled states experiments, the vector potential function $\vec{\alpha}_{k\lambda}(\vec{r}, t)$ with the quantized amplitude $\alpha_{0k}(\omega_k) = \xi\omega_k$ can play the role of a real wave function for the photon that can be suitably normalized. A k-mode photon as a quantum of the electromagnetic field and as an integral physical entity subsists over a period (thus over a wavelength) through the oscillation of the quantized vector potential at the angular frequency ω_k and propagates along the propagation axis according to the wave function $\vec{\alpha}_{k\lambda}(\vec{r}, t)$. At any point within a wavelength, the oscillation of the vector potential gives birth to intrinsic electric and magnetic fields, also oscillating at the same frequency, whose strength is proportional to the square of the angular frequency. This fundamental physical aspect needs to be further investigated experimentally. Following these results, non-destructive photon detection methods might be developed based on the detection of the influence of the photon intrinsic electric and/or magnetic fields upon the matter.

References

[1] A. Aspect *et al.*, Experimental tests of realistic local theories via Bell's theorem, *Phys. Rev. Lett.* 47, 7 (1981).

[2] A. Aspect *et al.*, Experimental test of Bell's inequalities using time-varying analyzers, *Phys. Rev. Lett.* 49, 25 (1982).

[3] G. Grynberg, A. Aspect, and C. Fabre, *Introduction to Quantum Optics* (Cambridge University Press, Cambridge, UK, 2010).

[4] W. Tittel *et al.*, Violation of Bell inequalities by photons more than 10 km apart, *Phys. Rev. Lett.* 81, 3563 (1998).

[5] G. Weihs *et al.*, Violation of Bell's inequality under strict Einstein locality conditions, *Phys. Rev. Lett.* 81, 5039 (1998).

[6] H. Zbinden *et al.*, Experimental test of nonlocal quantum correlation in relativistic configurations, *Phys. Rev. A* 63, 022111 (2001).

[7] M. A. Rowe *et al.*, Experimental violation of a Bell's inequality with efficient detection, *Nature* 409, 791 (2001).

[8] F. Matthew, Quantum entanglement results from quantum state transition at fast-than-light speed with matter wave's phase velocity, *Ars Technica* (30 October 2012).

[9] R. F. Werner, Quantum states with Einstein-Podolsky-Rosen correlations admitting a hidden-variable model, *Phys. Rev. A* 40, 4277 (1989).

[10] B. J. Smith and M. G. Raymer, Photon wave functions, wave-packet quantization of light, and coherence theory, *New J. Phys.* 9, 414 (2007).

[11] N. Chandrasekar, Quantum mechanics of photons, *Adv. Stud. Theor. Phys.* 6, 1 (2012).

[12] J. E. Sipe, Photon wave functions, *Phys. Rev. A* 52, 1875 (1995).

[13] D. L. Khokhlov, Spatial and temporal wave functions of photon, *Appl. Phys. Res.* 2, 1 (2010).

[14] C. Meis, Vector potential quantization and the quantum vacuum, *Phys. Res. Int.* 2014, 187432 (2014).

[15] C. Meis, Photon wave-particle duality and virtual electromagnetic waves, *Found. Phys.* 27, 865 (1997).

[16] C. Meis and P. R. Dahoo, Vector potential quantization and the photon wave-particle representation, *J. Phys.: Conf. Ser.* 738, 012099 (2016). DOI: 10.1088/1742-6596/738/1/012099.

[17] C. Meis, Electric potential of the quantum vacuum, *Phys. Essays* 22, 1 (2009).

[18] C. Meis, *Light and Vacuum*, 2nd ed. (World Scientific, Singapore, 2017).

[19] C. Meis, The vacuum energy density: QED beyond the standard description, *Int. Rev. Phys.* 10, 1 (2016).

[20] L. Mandel, Configuration-space photon number operators in quantum optics, *Phys. Rev.* 144, 1071 (1966).

[21] L. Mandel and E. Wolf, *Optical Coherence and Quantum Optics* (Cambridge University Press, Cambridge, UK, 1995).

[22] F. H. Read, *Electromagnetic Radiation* (John Wiley & Sons, New York, NY, USA, 1980).
[23] R. Loudon, *The Quantum Theory of Light*, 3rd ed. (Oxford University Press, Oxford, UK, 2000).
[24] J. C. Garrison and R. Y. Chiao, *Quantum Optics* (Oxford University Press, Oxford, UK, 2008).

Chapter 4

Quantized Vector Potential and the Photon Wave Function*

Abstract
The vector potential function $\vec{\alpha}_{k\lambda}(\vec{r},t)$ for a k-mode and λ-polarization photon, with the quantized amplitude $\alpha_{0k}(\omega_k) = \xi\omega_k$, satisfies the classical wave propagation equation as well as Schrodinger's equation with the relativistic massless Hamiltonian $\tilde{H} = -i\,\hbar\,c\vec{\nabla}$ and finally an equivalent quantum equation for the vector potential amplitude operator $\tilde{\alpha}_0 = -i\xi\,c\vec{\nabla}$. Thus, $\vec{\alpha}_{k\lambda}(\vec{r},t)$ behaves as a wave function for the photon within a non-local representation that can be suitably normalized. It is deduced that the probability for detecting a k-mode photon around a point on the propagation axis depends on the square of the angular frequency. Taking into account the left and right circularly polarized states and weighting $\vec{\alpha}_{k\lambda}(\vec{r},t)$ by $\omega_k\sqrt{2\varepsilon_0}$, we define a six-components wave function for a k-mode photon. The square of the modulus of the defined wave function gives the energy density at a given coordinate, which depends on the fourth power of the angular frequency. The amplitudes of the electric and magnetic fields of a single k-mode photon free of cavity are also calculated, and it is shown that they are proportional to the square of the angular frequency. In this way, the influence of the photon electric and/or magnetic fields on the energy levels of atoms and molecules might be used for a non-destructive photon detection.

*This chapter is reproduced from C. Meis and P. R. Dahoo. 2017. "Quantized vector potential and the photon wave function", *Journal of Physics: Conference Series*, 936, 012004.

1. Introduction

The challenging difficulties encountered in establishing a wave function for the photon are well developed in the literature [1–5]. Indeed, the photon as a relativistic massless particle constitutes a particular case, and a wave function within the quantum mechanical concept can hardly be defined. However, in recent years, the experimental evidence of the entangled states with single photon sources has demonstrated the permanent violation of Bell's inequality, entailing that the existence of hidden variables within a local quantum mechanical representation is excluded (see Ref. [6] and references therein). At the same time, it revealed the necessity of developing a non-local representation for the photon, through a real wave function, in which new variables are not explicitly excluded. Hence, we explore here the possibility of establishing a photon wave function through the vector potential with quantized amplitude. We also analyze the resulting expressions of the intrinsic electric and magnetic fields of a free photon, opening perspectives for new experiments that might be used for a non-destructive photon detection method.

2. Vector potential amplitude quantization

A detailed unit analysis of the general solution of Maxwell's equations for the vector potential has shown that it is proportional to the frequency. Consequently, for a k-mode photon with angular frequency ω_k, the vector potential amplitude α_{0k} can be written as [7–10]

$$\alpha_{0k}(\omega_k) = \xi\omega_k \tag{1}$$

where ξ is a constant.

In the first approximation, the value of the constant ξ has been evaluated [9, 11] by normalizing the mean energy of a plane electromagnetic wave over a period to Planck's expression for the energy of a single photon $\hbar\omega_k$, where $\hbar = h/2\pi$ is Planck's reduced constant, obtaining

$$|\xi| \propto \left| \frac{1}{(2\pi)^{3/2}} \sqrt{\frac{\hbar}{8\alpha\varepsilon_0 c^3}} \right| = \left| \frac{\hbar}{4\pi ec} \right| = 1.747\ \ 10^{-25}\ \mathrm{Vm^{-1}s^2} \tag{2}$$

with $\alpha = 1/137$ the fine structure constant, ε_0 the vacuum electric permittivity, c the speed of light in vacuum and e the electron-positron charge.

Hence, in the plane wave representation, the vector potential for a k-mode and λ-polarization photon can be written as

$$\vec{\alpha}_{k\lambda}(\vec{r},t) = \omega_k(\xi\hat{\varepsilon}_{k\lambda}e^{i(\vec{k}\cdot\vec{r}-\omega_k t+\theta)} + cc) = \omega_k\vec{\Xi}_{k\lambda}(\omega_k,\vec{r},t) \quad (3)$$

where cc is the complex conjugate and λ takes two values corresponding to the left (L) and right (R) circular polarization expressed by the unit vector ε_λ.

The general equation for the vector potential of an electromagnetic wave considered as a superposition of plane waves is now written as

$$\begin{aligned}\vec{A}(\vec{r},t) &= \sum_{k,\lambda}\omega_k\left[\xi\hat{\varepsilon}_{k\lambda}e^{i(\omega_k t-\vec{k}\cdot\vec{r}+\phi)} + \xi^*\hat{\varepsilon}^*_{k\lambda}e^{-i(\omega_k t-\vec{k}\cdot\vec{r}+\phi)}\right] \\ &= \sum_{k,\lambda}\omega_k\vec{\Xi}_{k\lambda}(\omega_k,\vec{r},t) \end{aligned} \quad (4)$$

According to the relation (1), the quantized vector potential amplitude is an intrinsic property of a single photon state, defining the vector potential function (3).

3. The vector potential as a wave function for the photon

It can be easily shown [7,9,12] that the vector potential function (3) satisfies the classical wave propagation equation in vacuum

$$\vec{\nabla}^2\vec{\alpha}_{k\lambda}(\vec{r},t) - \frac{1}{c^2}\frac{\partial^2}{\partial t^2}\vec{\alpha}_{k\lambda}(\vec{r},t) = 0 \quad (5)$$

as well as the *vector potential–energy* (*wave–particle*) equation

$$i\begin{pmatrix}\xi\\ \hbar\end{pmatrix}\frac{\partial}{\partial t}\vec{\alpha}_{k\lambda}(\vec{r},t) = \begin{pmatrix}\tilde{\alpha}_{0k}\\ \tilde{H}\end{pmatrix}\vec{\alpha}_{k\lambda}(\vec{r},t) \quad (6)$$

where the vector potential amplitude operator $\tilde{\alpha}_{0k} = -i\xi\, c\vec{\nabla}$ and the relativistic Hamiltonian for a massless particle $\tilde{H} = -i\,\hbar c\vec{\nabla}$ have the eigenvalues $\xi\omega_k$ and $\hbar\omega_k$, respectively [7–9].

Consequently, the vector potential function $\vec{\alpha}_{k\lambda}(\vec{r},t)$ with the quantized amplitude $\xi\omega_k$ behaves as a real wave function for the photon within a non-local representation. Obviously, when considering the propagation of a k-mode photon along, e.g., the z-axis, unlike the momentum operator $\tilde{P}_z$, which is well defined in quantum mechanics $\vec{P}_z = \hbar\vec{k}_z = \frac{h}{\lambda_k}\hat{e}_z \rightarrow \tilde{P}_z = -i\hbar\frac{\partial}{\partial z}$, it is not evident how to define a position operator, and this difficulty has been widely discussed in the literature [1,4,6,9,13]. In fact, Heisenberg's uncertainty for the position and momentum for a photon with a wavelength λ_k propagating along the z-axis can be written as

$$\delta z\, \delta P_z \geq h \;\rightarrow \delta z\, \delta\left(\frac{1}{\lambda_k}\right) \geq 1 \tag{7}$$

The physical meaning of Heisenberg's relation is that the photon as an integral entity cannot be localized precisely on a point of the propagation axis z except with a significant uncertainty, which is of the order of the wavelength λ_K. This fact is a natural consequence that lies in the intrinsic wave–particle nature of the photon [6,13, 15–17]. As quoted by many authors, despite the fact that the *point-particle* model used in the QED formalism is a practical mathematical tool, in reality, the photon cannot be defined within a length shorter than its wavelength. Thus, what we can essentially say is that when a k-mode photon is emitted at a coordinate z_e at an instant t_e and propagates in vacuum along the z-axis, the probability $P_k(z)$ to be localized at time t and at the coordinate $z = z_e + c(t - t_e)$ has rather to be considered within a volume having roughly the value λ_k^3 [6,16,17] around the point z, and according to (3), it is proportional to the square of the angular frequency

$$P_k(\vec{r}) \propto |\vec{\alpha}_{k\lambda}(\vec{r},t)|^2 \propto \xi^2\omega_k^2 \tag{8}$$

Considering the normalization factor $1/\xi\omega_{\text{max}}$ for $\vec{\alpha}_{k\lambda}(\vec{r},t)$ and taking $\omega_{\text{max}} \sim 10^{25}$ rad Hz, which corresponds to highly energetic gamma photons with wavelengths much shorter than the electron radius, then the detection probability becomes $P_k(\vec{r}) \propto \omega_k^2/\omega_{\text{max}}^2$ and tends to 1 for $\omega_k = \omega_{\text{max}}$. It is also worth noticing that, disregarding the physical units, Eq. (8) also tends naturally to one for the same value $\omega_k = \omega_{\text{max}} = 10^{25}$ rad Hz and can be practically used as a direct indication. Another normalization possibility consists of considering $\omega_{\text{max}} \sim 10^{43}$ rad Hz, corresponding roughly to Planck's energy

10^{19} GeV. However, we do not actually have any experimental evidence of photons with frequencies as high as 10^{43} Hz.

Now, taking into account the left (L) and right (R) circularly polarized states ($\lambda =$ L,R) and weighting the vector potential by the factor $\omega_k\sqrt{2\varepsilon_0}$, we define the six components wave function

$$\Phi_{k,(L,R)}(\vec{r},t) = \omega_k\sqrt{2\varepsilon_0}\begin{pmatrix}\vec{\alpha}_{k,L}(\vec{r},t)\\ \vec{\alpha}_{k,R}(\vec{r},t)\end{pmatrix} \tag{9}$$

Obviously, for both L and R polarizations, $\Phi_{k,(L,R)}(\vec{r},t)$ also satisfies the wave propagation equation (5) and the *vector potential–energy* equation (6). The square of the modulus of the defined wave function $|\Phi_{k,(L,R)}(\vec{r},t)|^2$ gives the energy density of the electromagnetic field composed of a single mode

$$W_k(\vec{r},t) = |\Phi_{k,(L,R)}(\vec{r},t)|^2 = 2\varepsilon_0\xi^2\omega_k^4 \tag{10}$$

It is worth noticing that the last expression depends on the fourth power of the angular frequency, as in the case of the energy density of a radiating dipole [13, 14].

We can also deduce the amplitude of the electric field $|\vec{\varepsilon}_k|$ of a free single k-mode photon [18], which can be expressed through the square of the angular frequency

$$|\vec{\varepsilon}_k| = \left|-\frac{\partial}{\partial t}\vec{\alpha}_{k\lambda}(\vec{r},t)\right| \propto \xi\omega_k^2 \tag{11}$$

Considering a plane wave representation, the corresponding magnetic field $|\vec{\beta}_k|$ of the photon is equally proportional to the square of the angular frequency

$$|\vec{\beta}_k| \propto \sqrt{\varepsilon_0\mu_0}\xi\omega_k^2 \tag{12}$$

where μ_0 is the vacuum magnetic permeability.

It is of fundamental interest to investigate experimentally the latest results, e.g., the coupling of left and right circularly polarized photons gives rise to a linearly polarized state whose spatial region around $\lambda_k/4$, where the electric field is maximum, corresponds (in the visible–UV range) to a considerable distance of many nanometers and can be crossed by a large number of atoms or molecules. Under these conditions, high-resolution laser spectroscopy might be employed for the detection of the Stark shifts induced by the photon electric field on the atomic or molecular energy levels. Such an experimental investigation could confirm the dependence of the single photon electric field on the square of the angular frequency. This might also open perspectives on the development of non-destructive photon detection through the measurement of the influence of the photon electric or magnetic fields upon the atomic or molecular energy levels.

4. Conclusion

We have seen that the vector potential function $\vec{\alpha}_{k\lambda}(\vec{r},t)$ with the quantized amplitude $\alpha_{0k}(\omega_k) = \xi\omega_k$ satisfies the propagation equation and the coupled *vector potential–energy* equation. It can therefore play the role of a real wave function for the photon in a non-local representation that can be suitably normalized. A k-mode photon as an integral physical entity subsists over a period (thus over a wavelength λ_k) through the oscillation of the quantized vector potential and propagates along the wave function. Weighting $\vec{\alpha}_{k\lambda}(\vec{r},t)$ by $\omega_k\sqrt{2\varepsilon_0}$, we may define a six-components wave function $\Phi_{k,(L,R)}(\vec{r},t)$ for the photon, including both circular polarizations. The square of the modulus of $\Phi_{k,(L,R)}(\vec{r},t)$ gives the energy density on a point of the propagation axis. The oscillation of the vector potential gives birth to intrinsic electric and magnetic fields within a single free photon state over a wavelength, which are proportional to the square of the frequency.

References

[1] I. Bialynicki-Birula, Photon wave function, in *Progress in Optics*, Vol. XXXVI, E. Wolf, Ed. (Elsevier, Amsterdam, The Netherlands, 1996).

[2] N. Chandrasekar, Quantum mechanics of photons, *Adv. Stud. Theor. Phys.* 6, 1 (2012).

[3] J. E. Sipe, Photon wave functions, *Phys. Rev. A.* 52, 1875 (1995).

[4] B. J. Smith and M. G. Raymer, Photon wave functions, wave packet quantization of light, and coherence theory, *New J. Phys.* 9, 414 (2007).

[5] D. L. Khokhlov, Spatial and temporal wave functions of photons, *Appl. Phys. Res.* 2, 1 (2010).

[6] G. Grynberg, A. Aspect, and C. Fabre, *Introduction to Quantum Optics* (Cambridge University Press, Cambridge, UK, 2010).

[7] C. Meis, Vector Vector potential quantization and the quantum vacuum, *Phys. Res. Int.* 2014, 187432 (2014).

[8] C. Meis, Photon wave-particle duality and virtual electromagnetic waves, *Found. Phys.* 27, 865 (1997).

[9] C. Meis, *Light and Vacuum*, 2nd ed. (World Scientific, Singapore, 2017).

[10] C. Meis, The vacuum energy density and the cosmological constant: QED beyond the standard description, *Int. Rev. Phys.* 10, 1 (2016).

[11] C. Meis, Electric potential of the quantum vacuum, *Phys. Essays* 22, 1 (2009).

[12] C. Meis and P. R. Dahoo, Vector potential quantization and the photon wave-particle representation, *J. Phys.: Conf. Ser.* 738, 012099 (2016). DOI: 10.1088/1742-6596/738/1/012099.

[13] J. C. Garrison and R. Y. Chiao, *Quantum Optics* (Oxford University Press, Oxford, UK, 2008).

[14] F. H. Read, *Electromagnetic Radiation* (John Wiley & Sons, New York, NY, USA, 1980).

[15] R. Loudon, *The Quantum Theory of Light*, 3rd ed. (Oxford University Press, Oxford, UK, 2000).

[16] L. Mandel and E. Wolf, *Optical Coherence and Quantum Optics* (Cambridge University Press, Cambridge, UK, 1995).

[17] L. Mandel, Configuration-space photon number operators in quantum optics, *Phys. Rev.* 144, 1071 (1966).

[18] C. Meis and P. R. Dahoo, Vector potential quantization and the photon intrinsic electromagnetic properties: Towards nondestructive photon detection, *Int. J. Quantum Inf.* 15(8), 1740003 (2017).

Chapter 5

Quantized Field of Single Photons*

Abstract
We present the theoretical developments describing the physical characteristics of a single photon in conformity with the experimental evidence. The quantization of the electromagnetic field vector potential amplitude is enhanced to a free-of-cavity photon state. Coupling the Schrödinger equation with the relativistic massless particle Hamiltonian to a symmetrical vector potential relation, we establish the *vector potential–energy* equation for the photon, expressing the simultaneous *wave–particle* nature of a single photon, an intrinsic physical property. It is shown that the vector potential can be naturally considered as a real wave function for the photon, entailing a coherent localization probability. We deduce directly the electric and magnetic field amplitudes of the cavity-free single photon, which are revealed to be proportional to the square of the angular frequency. The zero-energy electromagnetic field ground state (EFGS), a quantum vacuum real component, issues naturally from Maxwell's equations and the vector potential quantization procedure. The relation of the quantized amplitude of the photon vector potential to the electron–positron charge resulting directly shows the physical relationship between photons and electrons–positrons that might be at the origin of their mutual transformations. It becomes obvious that photons as well as electrons–positrons issue from the same quantum vacuum field.

---

*This chapter is reproduced from C. Meis. 2019. "Quantized Field of Single Photons", In: Keyu Xia, (Ed.), *Single Photon Manipulation*, IntechOpen, London, pp. 15–36.

1. Introduction

During the last decades, an impressive technological development has been achieved, permitting the manipulation of single photons with a high degree of statistical accuracy. However, despite the significant experimental advances, we still do not have a clear physical picture of a single photon state universally accepted by the scientific community, especially involved in quantum electrodynamics (QED). In this chapter, based on the present state of knowledge, we build a synthesis of the physical characteristics of a single photon put in evidence with the experiments and then advance to theoretical developments for its representation. Accordingly, the concept of the wave–particle nature of a single photon becomes physically comprehensive in agreement with the experimental evidence.

However, before advancing to the theoretical developments, we consider that it is important to start with a brief historical review of the efforts carried out previously for understanding the nature of light, while simultaneously building a synthesis of the main experimental results, which are of crucial importance for the comprehension of the birth of the photon concept.

The very first scientific publications on the nature of light are due to ancient Greeks who believed light to be composed of corpuscles [1,2]. Around 300 BC, Euclid published the book *Optica* in which he developed the laws of reflection based on the rectilinear propagation of light. Two centuries later, Ptolemy of Alexandria published the book *Optics*, in which he included extensively all the previous knowledge on light. In this book, colors as well as refraction of the moonlight and sunlight by the Earth's atmosphere were analyzed. Alhazen (965–1040 AD), the pioneer of the scientific study of modern optics, continued the works of ancient Greeks and demonstrated experimentally that light does not interact with light. After Alhazen, almost no progress was reported until the seventeenth century.

During the 1670s, Newton revived the ideas of ancient Greeks and advanced the theory, following which light is composed of corpuscles that travel rectilinearly [3]. Ten years later, Huygens developed the principles of the wave theory of light [1, 4, 5]. Huygens's wave theory was a hard opponent to Newton's corpuscular concept. At the beginning of the nineteenth century, Young obtained experimentally interference patterns using different sources of light and explained some polarization observations by assuming that light oscillations are perpendicular to the propagation axis [1, 6]. Euler and Fresnel explained the diffraction patterns observed experimentally by applying the wave theory [6]. In 1865, Maxwell published his theory on electromagnetic waves, establishing the relations between the electric and magnetic fields and showing that light is composed of electromagnetic waves [7]. A few years later, Hertz confirmed Maxwell's theory by discovering the long-wavelength electromagnetic radiation [1, 7]. Thus, at the end of the nineteenth century, the scientific community started to officially accept the wave nature of light, replacing Newton's theory.

Nevertheless, new events supporting the particle nature of light occurred at the beginning of the twentieth century. Stefan and Wien discovered the direct relationship between the thermal radiation energy and the temperature of a black body [8, 9]. However, the emitted radiation energy density as a function of the temperature calculated by Rayleigh failed to describe the experimental results at short wavelengths. Scientists had given the name "UV catastrophe" to this problem, revealing the necessity of a new theoretical approach. Planck managed to establish the correct energy density expression for the radiation emitted by a black body with respect to temperature, in excellent agreement with the experiment [8]. For this purpose, he assumed that the bodies are composed of "oscillators" which have the particularity of emitting the electromagnetic energy in "packets" of $h\nu$, where ν is the frequency and h is a constant that was later called Planck's constant. During the same period, the experiments carried out by Michelson and Morley [10] demonstrated that the speed of light in vacuum is a universal physical constant corresponding to the product of the frequency ν times the wavelength λ, i.e., $c = \lambda\nu$.

In 1902, Lenard pointed out that the photoelectric effect, discovered by Hertz 15 years earlier [11], occurs beyond a threshold frequency of light, and the kinetic energy of the emitted electrons does not depend on the incident light intensity. Based on Planck's works, Einstein proposed a simple interpretation of the photoelectric effect, assuming that the electromagnetic radiation is composed of quanta with energy $h\nu$ [12]. He advanced that *the energy of a light ray when spreading from a point consists of a finite number of energy quanta localized in points in space, which move without dividing and are only absorbed and emitted as a whole.* Although that was a decisive step toward the particle theory of light, the concept of the light quanta was still not generally accepted and Bohr, who was strongly opposed to the particle concept of light [13], announced in his Nobel lecture (1922) that *the light quanta hypothesis is not compatible with the interference phenomena and consequently it cannot throw light in the nature of radiation.* Bohr's statement was rather surprising because Taylor's experiments, consisting of repeating Young's double-slit diffraction at extremely low light intensities, had already demonstrated since 1909 that light rays are composed of discrete parts whose spots compose the diffraction patterns by gradual accumulation on the detection screen [14]. Compton published his studies on X-rays scattered by free electrons in 1923, advancing that the experimental results could only be interpreted based on the light quanta model [15]. Thus, the photoelectric effect and Compton scattering have been initially considered as the undoubtable demonstrations of the particle nature of light and historically were the strongest arguments in favor of the light quanta concept which started to be universally accepted, and Lewis introduced the word "photon," from the Greek *phos* ($\Phi\omega$s = Light) [1,4]. It is extremely important here to mention that Wentzel in 1926 [16], Beck in 1927 [17] as well as Lamb and Scully much later in the 1960s [18] demonstrated that the photoelectric effect can be interpreted remarkably well by only considering the wave nature of light without referring to photons at all [19]. Furthermore, the Compton scattering has been fully interpreted by Klein and Nishina in 1929 [20], also by considering the electromagnetic wave nature of light without invoking the photon concept. On the other hand, Young's experiment, initially presented as the most convincing argument for the wave nature of light, was applied by Taylor at very low intensities to demonstrate the particle concept of light [14]. Indeed, much later, Jin *et al.* [21] published an excellent theoretical interpretation of Young's diffraction experiments based only on the particle representation of light. Thus, the picture of the nature of light in the 1930s was rather confusing since both sides defending the wave or the particle nature advanced equally strong arguments. Hence, Bohr, inspired by De Broglie's thesis on the simultaneous wave character of particles, announced the *complementarity principle*, according to which *light has both wave and particle natures appearing mutually exclusively in each specific experimental condition* [1,19,22].

With the development of lasers [23] in the sixties and the revolutionary parametric down-converter techniques [24, 25] in the seventies, it became possible to realize conditions in which, with a convenient statistical confidence, only a single photon may be present in the experimental apparatus. In this way, the double prism experiment [26] realized in the nineties contradicted for the first time Bohr's *mutual exclusiveness,* demonstrating that a *single photon exhibits both the wave and particle natures in the same experimental conditions.*

According to the experimental investigations, it has always been stated that a photon has circular, left or right, polarization with spin $\pm h/2\pi$ and cannot be conceived along the propagation axis within a length shorter than its wavelength [27]. Indeed, since Mandel's experiments in the sixties [28, 29], all the efforts to precisely localize a single photon remained fruitless, and the general conclusion is that a photon cannot be better localized than within a volume of the order of the cube of its wavelength [30, 31]. Furthermore, Grangier *et al.* experimentally demonstrated the indivisibility of photons [19, 32], while in recent years, the entangled state experiments [30, 31] have shown that the photon should be locally an integral entity during the detection procedure, but with a real non-local wave function.

The lateral expansion of a single photon, considered locally as an indivisible entity, was always an intriguing part of physics. With the purpose of studying the lateral expansion of the electromagnetic rays, Robinson in 1953 [33] and Hadlock in 1958 [34] carried out experiments using microwaves crossing small apertures and deduced that no energy is transmitted through apertures, whose dimensions are smaller than roughly $\sim\lambda/4$. In 1986, for the same purpose, Hunter and Wadlinger [35, 36] used X-band microwaves with $\lambda = 28.5\,\text{mm}$ and measured the transmitted power through rectangular or circular apertures of different dimensions. They concluded that no energy is transmitted when the apertures are smaller than $\sim\lambda/\pi$, confirming that the lateral expansion of the photons is a fraction of the wavelength.

Thus, the experiments have shown that the single photon is not a point and cannot be localized at a coordinate, as stated by Einstein, while it can exhibit both the wave and particle natures in the same experimental conditions, contradicting Bohr's mutual exclusiveness. However, QED was developed during the 1930s–1960s based upon the point particle model for the photon [37–39]. In fact, the point photon concept has permitted us to establish an efficient mathematical approach for describing states before and after an interaction process [19, 40–42], but it is naturally inappropriate for the description of the real nature of a single photon.

Finally, what we can essentially draw out by summing up the experimental evidence is that a single photon is a minimum, local, indivisible part of the electromagnetic field with precise energy $h\nu$, momentum $h\nu/c$, having circular left or right polarization with spin $\pm h/2\pi$, respectively. It is not a point particle since it expands over a wavelength λ along the propagation axis and is detectable within a volume of the order of λ^3, implying that its lateral expansion is a fraction of its wavelength. Hence, it appears to be a local "wave–corpuscle" guided by a non-local wave function, absorbed and emitted as a whole and capable of interacting with charges increasing or decreasing its frequency and consequently its energy.

In the following sections, we present first the standard theoretical representation of the electromagnetic field quantization resulting in photons, and next, we proceed to recent advances based on the vector potential quantization enhanced to a single photon state.

2. The electromagnetic field vector potential

2.1. *Reality of the vector potential*

Since the formulation of Maxwell's equations, the vector potential $\vec{A}(\vec{r},t)$ has been considered as a pure mathematical artifact [4, 5, 7] used to calculate the electric field

$$\vec{E}(\vec{r},t) = -\partial\vec{A}(\vec{r},t)/\partial t - \vec{\nabla}\Phi(\vec{r},t) \tag{1}$$

where $\Phi(\vec{r},t)$ is the scalar potential as well as the magnetic field

$$\vec{B}(\vec{r},t) = \vec{\nabla}\times\vec{A}(\vec{r},t) \tag{2}$$

In 1949, Ehrenberg and Siday were the first to put in evidence the influence of the vector potential on charged particles [43], deducing that it is a real physical field. Ten years later, Aharonov and Bohm reaffirmed the influence of the vector potential on electrons in the complete absence of electric and magnetic fields [44]. This was confirmed experimentally by Chambers [45], Tonomura [46] and Osakabe [47], demonstrating without any doubt the reality of the vector potential and its direct influence on charges.

From a theoretical point of view [44], the behavior of a particle with charge q and mass m in the vicinity of a solenoid where the vector potential is present is described by the Hamiltonian

$$H = \frac{1}{2m}(-i\hbar\vec{\nabla} - q\vec{A}(\vec{r},t))^2 + q\Phi(\vec{r},t) \tag{3}$$

with $\hbar = h/2\pi$ Planck's reduced constant.

If the solenoid is extremely long along the z-axis, then the magnetic field is uniform in the region inside and zero outside. The scalar potential Φ can be put to zero by assuming that the solenoid is not charged. In this case, in the outside region, the electric and magnetic fields are zero, but the vector potential is not zero and depends on the magnetic field flux in the solenoid

$$\vec{A} = \frac{1}{2\pi\, r}\int_s \vec{B}\cdot d\vec{S}\,\hat{e}_\theta \tag{4}$$

where r is the radial distance from the z-axis of the solenoid, S is the surface of the circle with radius r perpendicular to z and $\hat{e}_\theta$ the angular unit vector in cylindrical coordinates.

The Schrödinger equation for a charged particle outside the solenoid, where the vector potential is not zero, is written in the complete absence of any other external potential

$$i\hbar\frac{\partial}{\partial t}\Psi(\vec{r},t) = \frac{1}{2m}(i\hbar\vec{\nabla} + q\vec{A})^2\Psi(\vec{r},t) \tag{5}$$

with $\vec{A}$ given by Eq. (4). The solutions of the last equation are the wave functions

$$\Psi(\vec{r},t) = \Psi_p(\vec{r},t)\, e^{\frac{iq}{\hbar}\int_0^{\vec{r}}\vec{A}(\vec{r}')\cdot d\vec{r}'} \tag{6}$$

where $\Psi_p(\vec{r},t)$ is the solution of Schrödinger's equation in the absence of the vector potential

$$i\hbar\frac{\partial}{\partial t}\Psi_p(\vec{r},t) = -\frac{\hbar^2}{2m}\vec{\nabla}^2\,\Psi_p(\vec{r},t) \tag{7}$$

The exponential part of the wave function of Eq. (6) entails that two particles having equal charge and mass, moving both outside the solenoid at the same distance from the axis, but the first in the same direction with the vector potential $\vec{A}$ and the second in the opposite direction will suffer a phase difference

$$\delta\Theta = \frac{q}{\hbar}\int_s \vec{B}\cdot d\vec{S} \tag{8}$$

Interference patterns for electrons in analogue conditions have been observed experimentally [45–47], demonstrating that the vector potential is a real physical field and interacts directly with charged particles in the complete absence of magnetic and electric fields and of any other potential.

2.2. *The radiation vector potential: Classical to quantum link*

The vector potential, being a real field, is considered as the fundamental link between the electromagnetic wave theory issued from Maxwell's equations and QED [19, 38, 40]. We will show analytically how this link is established.

In the classical theory [5, 7], the energy density of a mode k of the electromagnetic wave is written as

$$W_k(\vec{r},t) = \frac{1}{2}\left(\varepsilon_0|\vec{E}_k(\vec{r},t)|^2 + \frac{1}{\mu_0}|\vec{B}_k(\vec{r},t)|^2\right) \tag{9}$$

with ε_0 and μ_0 the electric permittivity and magnetic permeability of the vacuum, respectively, related to the speed of light in vacuum c by $\varepsilon_0\mu_0 c^2 = 1$.

In the case of a monochromatic plane wave with angular frequency ω_k, the electric $\vec{E}_k(\vec{r},t)$ and magnetic $\vec{B}_k(\vec{r},t)$ fields are proportional to the vector potential amplitude $A_{0k}(\omega_k)$:

$$\vec{E}_k(\vec{r},t) = -2\omega_k\, A_{0k}(\omega_k)\,\hat{\varepsilon}\,\sin(\vec{k}\cdot\vec{r} - \omega_k\, t) \tag{10}$$

$$\vec{B}_k(\vec{r},t) = -\frac{1}{c}2\,\omega_k\, A_{0k}(\omega_k)\,(\hat{k}\times\hat{\varepsilon})\,\sin(\vec{k}\cdot\vec{r} - \omega_k t) \tag{11}$$

where $\widehat{\varepsilon}$ is a unit vector perpendicular to the propagation axis, $|\vec{k}| = 2\pi/\lambda_k$ is the wave vector along the propagation axis and λ_k is the wavelength of the mode k.

Introducing Eqs. (10) and (11) in Eq. (9), the energy density now depends on the square of the vector potential amplitude

$$W_k(\vec{r},t) = 4\varepsilon_0\omega_k^2|A_{0k}(\omega_k)|^2 \sin^2(\vec{k}\cdot\vec{r} - \omega_k\, t) \tag{12}$$

The mean value over a period, and thus over a wavelength, is time-independent:

$$W_k = 2\varepsilon_0\omega_k^2|A_{0k}(\omega_k)|^2 \tag{13}$$

Note that the last equation expressing the mean energy density over a period of the mode k of the electromagnetic wave is independent of any external volume, yielding that in the classical description, a free-of-cavity electromagnetic radiation mode expands naturally within a minimum volume. In a given cavity, this volume corresponds roughly to that imposed by the boundary conditions and the cut-off wave vectors [4, 5, 7].

On the other hand, in the quantum description, the energy density for a number $N(\omega_k)$ of k-mode photons with angular frequency ω_k and energy $\hbar\omega_k$ in a volume V is written simply as

$$W_k = \frac{N(\omega_k)\hbar\omega_k}{V} \tag{14}$$

In order to link the classical to the quantum description [4,9,19], the classical mean energy density over a period, expressed by Eq. (13), is imposed to be equivalent to the quantum mechanics expression of Eq. (14) for $N(\omega_k) = 1$, obtaining the vector potential amplitude for a single k-mode photon:

$$|A_{0_k}(\omega_k)| = \sqrt{\frac{\hbar}{2\varepsilon_0\omega_k\, V}} \tag{15}$$

The last relation is the fundamental link between the classical and quantum theory of light, which is used to define in QED the vector potential amplitude operators for a single photon [19, 27, 30, 37–42]:

$$\tilde{A}_{k\lambda} = \sqrt{\frac{\hbar}{2\varepsilon_0\omega_k V}}a_{k\lambda} \quad \tilde{A}^*_{k\lambda} = \sqrt{\frac{\hbar}{2\varepsilon_0\omega_k V}}a^+_{k\lambda} \tag{16}$$

where $a_{k\lambda}$ and $a^+_{k\lambda}$ are, respectively, the annihilation and creation non-Hermitian operators for a k-mode and λ-polarization photon.

Here, it is worth noting that an external arbitrary volume parameter V appears in the vector potential amplitude of the single photon, expressed by Eq. (15), which is supposed to be an intrinsic physical property. This could entail the unphysical interpretation that a single photon in an infinite cavity has zero vector potential, thus, zero electric and magnetic fields and consequently zero energy.

This ambiguity, which is scarcely quoted in the literature, is lifted by considering that, in the case of a single photon, the volume V in Eq. (15) is equivalent to that defined by the boundary conditions in a cavity for the radiation mode k composed of a single photon.

3. Electromagnetic field quantization and the photon description

3.1. *Harmonic oscillator representation of the electromagnetic field*

The energy of the electromagnetic field in a volume V considered as a superposition of different k-modes and λ-polarizations is obtained directly from Eq. (13):

$$E_{EM} = 2\varepsilon_0 V \sum_{k,\lambda} \omega_k^2 |A_{k\lambda}(\omega_k)|^2 \tag{17}$$

where the summation over the polarizations λ takes only two values corresponding to circular left and right [19, 37–42].

Replacing in Eq. (17) the vector potential amplitude and its conjugate by the relations of the vector potential amplitude operators defined in Eq. (16), we get the "normal ordering" Hamiltonian corresponding to the order $a_{k\lambda}^{+} a_{k\lambda}$ of the creation and annihilation operators

$$\tilde{H}_{NO} = \sum_{k,\lambda} a_{k\lambda}^{+} a_{k\lambda} \hbar\omega_k \tag{18}$$

and the "anti-normal ordering" Hamiltonian corresponding to the order $a_{k\lambda} a_{k\lambda}^{+}$:

$$\tilde{H}_{ANO} = \sum_{k,\lambda} (a_{k\lambda}^{+} a_{k\lambda} + 1)\hbar\omega_k \tag{19}$$

where we have used the fundamental commutation relation in QED

$$[a_{k\lambda}, a^{+}_{k'\lambda'}] = \delta_{kk'}\delta_{\lambda\lambda'} \tag{20}$$

In Dirac's representation, the eigenfunctions take the simple expression $|n_{k\lambda}\rangle$, and the action of the creation and annihilation operators of a single k-mode and λ-polarization photon are written as

$$a^{+}_{k\lambda}|n_{k\lambda}\rangle = \sqrt{n_{k\lambda}+1}|n_{k\lambda}+1\rangle; \quad a_{k\lambda}|n_{k\lambda}\rangle = \sqrt{n_{k\lambda}}|n_{k\lambda}-1\rangle \tag{21}$$

The successive action of both operators in the normal order corresponds to the photon number Hermitian operator $\tilde{n}_{k\lambda} = a^{+}_{k\lambda}a_{k\lambda}$ having the eigenvalue $n_{k\lambda}$ representing the number of k-mode and λ-polarization photons

$$a^{+}_{k\lambda}a_{k\lambda}|n_{k\lambda}\rangle = \tilde{n}_{k\lambda}|n_{k\lambda}\rangle = n_{k\lambda}|k\lambda\rangle \tag{22}$$

In this representation, the normal and anti-normal ordering Hamiltonians are written, respectively, as

$$\tilde{H}_{NO} = \sum_{k,\lambda} \tilde{n}_{k\lambda}\hbar\omega_k; \quad H_{ANO} = \sum_{k,\lambda} (\tilde{n}_{k\lambda}+1)\hbar\omega_k \tag{23}$$

We obtain a harmonic oscillator Hamiltonian for the electromagnetic field by considering the mean value of the normal ordering and anti-normal ordering Hamiltonians

$$\tilde{H}_{EM} = \frac{1}{2}(\tilde{H}_{NO} + \tilde{H}_{ANO}) = \sum_{k,\lambda}\left(\tilde{n}_{k,\lambda} + \frac{1}{2}\right)\hbar\omega_k \tag{24}$$

Thus, in QED, the electromagnetic field is considered to be an ensemble of harmonic oscillators, each represented by a point particle, the photon, whose eigenfunction is denoted simply by $|1\rangle$ [19, 40, 42].

Although we have no experimental facts showing the harmonic oscillator nature of a single photon, this representation has been adopted since the 1930s [37].

In a different way, a harmonic oscillator representation for the electromagnetic field can be obtained by the intermediate of the canonical variables of position $Q_{k\lambda}$ and momentum $P_{k\lambda}$. For that purpose, we introduce the definitions expressing the vector potential amplitude and its complex conjugate with respect to $Q_{k\lambda}$ and $P_{k\lambda}$ [19, 30, 42]:

$$A_{k\lambda} = \frac{(\omega_k Q_{k\lambda} + iP_{k\lambda})}{2\omega_k\sqrt{\varepsilon_0 V}}; \quad A^*_{k\lambda} = \frac{(\omega_k Q_{k\lambda} - iP_{k\lambda})}{2\omega_k\sqrt{\varepsilon_0 V}} \tag{25}$$

Introducing the last expressions in Eq. (17), we get the electromagnetic field energy

$$E_{EM} = \frac{1}{2}\sum_{k,\lambda}(P^2_{k\lambda} + \omega_k^2 Q^2_{k\lambda}) \pm i\omega_k[Q_{k\lambda}, P_{k\lambda}] \tag{26}$$

where the (+) sign is obtained when Eq. (17) is considered initially to be in the "normal order", $A^*_{k\lambda}A_{k\lambda}$, and the (−) one when in the "anti-normal order" $A_{k\lambda}A^*_{k\lambda}$.

With the purpose of establishing a harmonic oscillator representation for the electromagnetic field, it is generally considered that $[Q_{k\lambda}, P_{k\lambda}] = 0$ in Eq. (26) because $Q_{k\lambda}$ and $P_{k\lambda}$ are simply canonical variables, obtaining the energy of an ensemble of harmonic oscillators:

$$E_{EM} = \frac{1}{2}\sum_{k,\lambda}(P^2_{k\lambda} + \omega_k^2 Q^2_{k\lambda}) \tag{27}$$

Replacing in the last equation the classical canonical variables of position and momentum with the corresponding Hermitian operators [19, 30, 42]

$$\tilde{P}_{k\lambda} = i\sqrt{\frac{\hbar\omega_k}{2}}(a^+_{k\lambda} - a_{k\lambda}); \quad \tilde{Q}_{k\lambda} = \sqrt{\frac{\hbar}{2\omega_k}}(a^+_{k\lambda} + a_{k\lambda}) \tag{28}$$

and putting $a^+_{k\lambda}a_{k\lambda} = \tilde{n}_{k\lambda}$, one gets the harmonic oscillator Hamiltonian for the radiation field

$$\tilde{H}_{EM} = \sum_{k,\lambda}\left(\tilde{n}_{k,\lambda} + \frac{1}{2}\right)\hbar\omega_k \tag{29}$$

At that level, it is important to note that, for a harmonic oscillator of a particle with mass m and momentum $\vec{p} = m\,d\vec{q}/dt$, with canonical variables of position $Q = |\vec{q}|\sqrt{m}$ and momentum $P = |\vec{p}|/\sqrt{m}$, the transition from the classical expression of energy

$$E = \frac{1}{2}(P^2 + \omega^2 Q^2) \tag{30}$$

to the quantum mechanics Hamiltonian

$$\tilde{H} = \frac{1}{2}(\tilde{P}^2 + \omega^2 \tilde{Q}^2) = \left(a^+ a + \frac{1}{2}\right)\hbar\omega = \left(\tilde{n} + \frac{1}{2}\right)\hbar\omega \tag{31}$$

where $\tilde{P} = i\sqrt{\frac{\hbar\omega}{2}}(a^+ - a)$, $\tilde{Q} = \sqrt{\frac{\hbar}{2\omega}}(a^+ + a)$ and $\tilde{n} = a^+ a$ is direct and needs no commutation operations between the canonical variables P and Q [19, 40].

Consequently, the harmonic oscillator Hamiltonian for a particle of mass m expressed by Eq. (31) is quite a physical result (e.g., phonons in solid state physics) obtained with a perfect correspondence between the classical canonical variables of momentum and position P and Q, respectively, and the corresponding Hermitian operators $\tilde{P}$ and $\tilde{Q}$.

Conversely, this is not the case for the electromagnetic field [19, 30, 40] because commutations between the canonical variables $Q_{k\lambda}$ and $P_{k\lambda}$ occur during the mathematical transition from Eq. (17) to Eq. (26). It is then considered that $[Q_{k\lambda}, P_{k\lambda}] = 0$ in order to obtain Eq. (27), just before replacing the classical canonical variables by the corresponding quantum operators. Here, it is important to remark that Heisenberg's commutation relation $[\tilde{Q}_{k\lambda}, \tilde{P}_{k'\lambda'}] = i\hbar\delta_{kk'}\delta_{\lambda\lambda'}$ is a fundamental concept of quantum mechanics, which should not be ignored when replacing classical variables by the corresponding quantum mechanics operators [19]. In fact, without dropping $[Q_{k\lambda}, P_{k\lambda}]$ in Eq. (26) and replacing the canonical variables by the corresponding quantum operators of Eq. (28), we get naturally the same normal ordering and anti-normal ordering Hamiltonians as in Eq. (23):

$$\tilde{H}^{(+)}_{EM} = \frac{1}{2}\sum_{k,\lambda}(\tilde{P}^2_{k\lambda} + \omega_k^2 \tilde{Q}^2_{k\lambda}) + i\omega_k[\tilde{Q}_{k\lambda}, \hat{P}_{k'\lambda'}]$$

$$\rightarrow \tilde{H}_{NO} = \sum_{k,\lambda} \tilde{n}_{k\lambda}\,\hbar\omega_k$$

$$\tilde{H}_{EM}^{(-)} = \frac{1}{2}\sum_{k,\lambda}(\tilde{P}_{k\lambda}^2 + \omega_k^2\tilde{Q}_{k\lambda}^2) - i\omega_k[\tilde{Q}_{k\lambda}, \tilde{P}_{k'\lambda'}]$$

$$\rightarrow \tilde{H}_{ANO} = \sum_{k,\lambda}(\tilde{n}_{k\lambda} + 1)\ \hbar\omega_k \tag{32}$$

Obviously, as frequently quoted [2, 19, 40], the fundamental mathematical ambiguity consisting of canceling the commuting classical variables term $[Q_{k\lambda}, P_{k\lambda}] = 0$ before the substitution by non-commuting quantum mechanics operators $[\tilde{Q}_{k\lambda}, \tilde{P}_{k'\lambda'}] = i\hbar\delta_{kk'}\delta_{\lambda\lambda'}$ leads to the harmonic oscillator Hamiltonian for the electromagnetic field.

In fact, since no experiment has yet demonstrated that a single photon is a harmonic oscillator, the main reason for considering the electromagnetic field as an ensemble of harmonic oscillators lies in the importance of the zero-point energy (ZPE) issued in the absence of photons from the eigenvalue $n_{k\lambda} = 0$ of Eq. (29) corresponding to the vacuum energy

$$E_{ZPE} = \sum_{k,\lambda}\frac{1}{2}\hbar\omega_k \tag{33}$$

The summation of the last expression over all modes and polarizations is infinite and represents the principal singularity in the QED formalism [19, 22, 27, 30, 40].

Nevertheless, the zero-point energy is very important because it is considered to be the basis for the explanation of the vacuum effects, such as the spontaneous emission, the Lamb shift and the Casimir effect. However, as pointed out by many authors [19, 27, 40, 42], it is worth underlining that the explanation of the spontaneous emission and the Lamb shift in QED is not due to Eq. (33) but precisely to the commutation properties of the photon creation and annihilation operators, $a_{k\lambda}^+$ and $a_{k\lambda}$, respectively. It has to be emphasized that in quantum theory, Eq. (33), being a constant, commutes with all quantum mechanics Hermitian operators corresponding to physical observables and consequently has absolutely no influence on any quantum process.

Conversely, the zero-point energy expressed by Eq. (33) is useful for the explanation of the spontaneous emission and the Lamb shift in the classical description of radiation [38, 40, 48].

Regarding the Casimir effect, it is often commented that caution has to be taken concerning the interpretation of its physical origin because it has been demonstrated by different methods [49–51] that it can be easily explained without invoking at all the zero-point energy.

Hence, in view of the above, the normal ordering Hamiltonian is the one mainly used in QED, casting aside the vacuum singularity issued from the harmonic oscillator formalism, while the zero-point energy issued from the harmonic oscillator Hamiltonian is principally useful in the classical formalism for the interpretation of the vacuum effects [19, 30, 42, 52].

3.2. *Electromagnetic field vector potential quantization in QED*

We have analysed in Section 3.1 the electromagnetic field energy quantization according to the harmonic oscillator representation. Now, we will analyse the vector potential field quantization following the second quantization process.

Considering the natural units ($\hbar = c = 1$), the radiation vector potential is written within the frame of the quantum field theory (QFT) [27, 39] as

$$\vec{A}(x) = \int \frac{d^3k}{(2\pi)^3 2k_0} \sum_{\lambda=1}^{2} \hat{\varepsilon}_{k\lambda} \left[\alpha^{(\lambda)}(k) e^{-ikx} + \alpha^{(\lambda)+}(k) e^{ikx}\right] \tag{34}$$

with

$$\left[\alpha^{(\lambda)}(k), \alpha^{(\lambda')+}(k')\right] = 2k_0 (2\pi)^3 \delta_{\lambda\lambda'} \delta^3(\vec{k} - \vec{k}') \tag{35}$$

where $\delta_{\lambda\lambda'}$ is the Kronecker delta, $\delta^3(\vec{k} - \vec{k}')$ is the Dirac delta function and

$$\alpha^{(\lambda)}(k) = (2k_0)^{1/2} (2\pi)^{3/2} \alpha_{k\lambda}; \quad \alpha^{(\lambda)+}(k) = (2k_0)^{1/2} (2\pi)^{3/2} \alpha^{+}_{k\lambda} \tag{36}$$

Using Eq. (36) in Eq. (34), the vector potential becomes

$$\vec{A}(x) = \int \frac{d^3k}{(2\pi)^{3/2} (2k_0)^{1/2}} \sum_{\lambda=1}^{2} \hat{\varepsilon}_{k\lambda} \left[\alpha_{k\lambda} e^{-ikx} + \alpha^{+}_{k\lambda} e^{ikx}\right] \tag{37}$$

with $\omega_k = (\vec{k}^2 + l^2)^{1/2}$ and $k^2 = k_0^2 - \vec{k}^2 = l^2$.

For $k = l = 0$, then $k_0 = |\vec{k}| = \omega_k$, and Eq. (37) is written as

$$\vec{A}(x) = \int \frac{d^3k}{(2\pi)^{3/2}} \sqrt{\frac{1}{2\omega_k}} \sum_{\lambda=1}^{2} \hat{\varepsilon}_{k\lambda} \left[\alpha_{k\lambda} e^{-ikx} + \alpha_{k\lambda}^{+} e^{ikx} \right] \tag{38}$$

Suppressing the natural units (i.e., introducing c and $\hbar$) and transforming the last equation in the SI system, which is generally used in QED, we get [39, 42, 52]

$$\vec{A}(x) = \int \frac{d^3k}{(2\pi)^{3/2}} \sqrt{\frac{\hbar}{2\varepsilon_0\omega_k}} \sum_{\lambda=1}^{2} \hat{\varepsilon}_{k\lambda} \left[\alpha_{k\lambda} e^{-ikx} + \alpha_{k\lambda}^{+}(k) e^{ikx} \right] \tag{39}$$

On the basis of the density of states theory, the quantization of a field in a cavity of volume V permits us to transform the continuous summation over the modes to a discrete one [19, 52]

$$\int \frac{d^3k}{(2\pi)^{3/2}} = \sqrt{\frac{1}{V}} \sum_k \tag{40}$$

The last transformation is only valid for an ensemble of k-modes whose wavelengths λ_k are much shorter than the characteristic dimensions of the volume V [19, 30, 40, 42].

Switching now to the Heisenberg representation,

$$\alpha_{k\lambda}(t) = \alpha_{k\lambda} e^{-i\omega_k t}; \quad \alpha_{k\lambda}^{+}(t) = \alpha_{k\lambda}^{+} e^{i\omega_k t} \tag{41}$$

generalizing the coordinate system and using Eq. (40), the vector potential of the electromagnetic field can be written in QED as [19, 30, 40, 42, 52]

$$\vec{A}(\vec{r}, t) = \sum_k \sqrt{\frac{\hbar}{2\varepsilon_0\omega_k V}} \sum_{\lambda=1}^{2} \left[\alpha_{k\lambda} \hat{\varepsilon}_{k\lambda} e^{i(\vec{k}\cdot\vec{r} - \omega_k t)} + \alpha_{k\lambda}^{+} \hat{\varepsilon}_{k\lambda}^{*} e^{-i(\vec{k}\cdot\vec{r} - \omega_k t)} \right] \tag{42}$$

Considering the scalar potential to be constant, the electric field is

$$\vec{E}(\vec{r}, t) = i \sum_k \sqrt{\frac{\hbar\omega_k}{2\varepsilon_0 V}} \sum_{\lambda=1}^{2} \left[\alpha_{k\lambda} \hat{\varepsilon}_{k\lambda} e^{i(\vec{k}\cdot\vec{r} - \omega_k t)} - \alpha_{k\lambda}^{+} \hat{\varepsilon}_{k\lambda}^{*} e^{-i(\vec{k}\cdot\vec{r} - \omega_k t)} \right] \tag{43}$$

The last expressions represent in a given volume V the quantized vector potential and the electric field of the electromagnetic radiation composed of a large number of modes k, each with angular frequency ω_k and wavelength $\lambda_k = 2\pi c/\omega_k$ much smaller than $V^{1/3}$:

$$\lambda_k \ll V^{1/3} \quad (\forall k) \tag{44}$$

The amplitudes in Eqs. (42) and (43) have been obtained using the density of states theory and are valid only on the condition of Eq. (44). Furthermore, the boundary conditions of the electromagnetic waves considered in cavities and waveguides impose the wave vectors k of the modes to be higher than a characteristic cut-off value $k > k_{\text{cut-off}}$ ($\lambda_k < \lambda_{\text{cut-off}}$) depending on the dimensions and the shape of the volume containing the radiation field [4–7]. Consequently, for a volume V with finite dimensions, the summation in Eqs. (42) and (43) runs only over the modes k with wave vectors higher than the minimum cut-off value $k_{\text{cut-off}}(V)$ corresponding to the shape and dimensions of V, so that we can write more precisely

$$\vec{A}(\vec{r},t) = \sum_{k>k_{\text{cut-off}}(V)} \sqrt{\frac{\hbar}{2\varepsilon_0\omega_k V}} \times \sum_{\lambda=1}^{2} \left[\alpha_{k\lambda}\hat{\varepsilon}_{k\lambda}e^{i(\vec{k}\cdot\vec{r}-\omega_k t)} + \alpha_{k\lambda}^{+}\hat{\varepsilon}_{k\lambda}^{*}e^{-i(\vec{k}\cdot\vec{r}-\omega_k t)}\right] \tag{45}$$

$$\vec{E}(\vec{r},t) = i \sum_{k>k_{\text{cut-off}}(V)} \sqrt{\frac{\hbar\omega_k}{2\varepsilon_0 V}} \times \sum_{\lambda=1}^{2} \left[\alpha_{k\lambda}\hat{\varepsilon}_{k\lambda}e^{i(\vec{k}\cdot\vec{r}-\omega_k t)} - \alpha_{k\lambda}^{+}\hat{\varepsilon}_{k\lambda}^{*}e^{-i(\vec{k}\cdot\vec{r}-\omega_k t)}\right] \tag{46}$$

The last equations represent the vector potential and the electric field of a large number of modes k of the quantized electromagnetic field in a finite volume V with $\lambda_k \ll V^{1/3}$ $(\forall k)$.

4. Quantized vector potential and photon representation

We have seen in Section 3.1 that, according to the energy quantization procedure, the photon is considered to be a point harmonic oscillator represented by the simplified eigenfunction $|1\rangle$. On the other hand, following the field quantization procedure in Section 3.2, it appears clearly that the established vector potential and electric field expressions in Eqs. (42) and (43), under the condition of Eq. (44), concern a large number of photons in a considerably large volume compared to their wavelengths. Thus, with the aim of obtaining a clearer picture of the single photon, we will now complement the previous descriptions by enhancing the vector potential amplitude quantization to the photon level.

4.1. *Photon vector potential amplitude and quantization volume*

As mentioned in Section 2.2, the classical expression of the mean energy density over a period for a single electromagnetic mode k, represented by Eq. (13), can be considered equivalent to that for a single photon in the quantum representation, given by Eq. (14) for $N(\omega_k) = 1$, on the condition that the volume V is not arbitrary and corresponds to that defined by the boundary conditions in a cavity for the considered electromagnetic mode. In fact, the physical properties of a free photon are independent of any surrounding volume unless the characteristic dimensions of the last one are of the order of the photon wavelength [53].

We recall again that the experimental evidence has shown [19, 28, 30, 52] that a single photon with angular frequency ω_k and wavelength $\lambda_k = 2\pi c/\omega_k$ can only be localized within a volume V_k whose dimensions are roughly the cube of its wavelength

$$V_k \propto \lambda_k^3 \quad \Rightarrow \quad V_k \propto \omega_k^{-3} \tag{47}$$

From a theoretical point of view, this is also compatible with the density of states theory according to which the spatial volume corresponding to a single state of the quantized field is proportional to ω_k^{-3} [19, 30, 40, 42].

On the other hand, the dimension analysis of the vector potential issued from the general solution of Maxwell's equations yields that it is proportional to an angular frequency [5, 7, 9]

$$\vec{A}(\vec{r}, t) = \frac{\mu}{4\pi} \int \frac{\vec{J}\left(\vec{r'}, t - \frac{|\vec{r} - \vec{r'}|}{c}\right)}{|\vec{r} - \vec{r'}|} d^3r' \quad \propto \quad \omega \tag{48}$$

where J is the current density $(Cm^{-2}s^{-1})$.

Indeed, it is well established experimentally that the energy density radiated by a dipole is proportional to ω^4, entailing from Eq. (12) that the vector potential amplitude is normally proportional to ω [4, 5, 7].

This result is gauge-independent since it concerns the natural units of the vector potential.

According to the previous considerations, for a free single k-mode photon with λ-polarization (left or right circular), the vector potential can be written in quantum and classical formalisms as

$$\tilde{\alpha}_{k\lambda}(\vec{r},t) = \alpha_{0k}(\omega_k)\left[\hat{\varepsilon}_{k\lambda}\alpha_{k\lambda}e^{i(\vec{k}\cdot\vec{r}-\omega_k t)} + \hat{\varepsilon}^*_{k\lambda}\alpha^+_{k\lambda}e^{-i(\vec{k}\cdot\vec{r}-\omega_k t)}\right]$$
$$\vec{\alpha}_{k\lambda}(\vec{r},t) = \alpha_{0k}(\omega_k)\left[\hat{\varepsilon}_{k\lambda}e^{i(\vec{k}\cdot\vec{r}-\omega_k t)} + \hat{\varepsilon}^*_{k\lambda}e^{-i(\vec{k}\cdot\vec{r}-\omega_k t)}\right] \quad (49)$$

where, following the above analysis, the amplitude is written as

$$\alpha_{0k}(\omega_k) = \xi\omega_k \quad (50)$$

with ξ being a constant [2, 54–56].

We can evaluate ξ [2, 54] by using Eqs. (49) and (50) in Eq. (13) and normalizing the energy to that of a single photon, $\hbar\omega_k$, by integrating over a wavelength along the propagation direction while taking into account the experimental results on the lateral expansion of the photon [33–36, 58]. We get

$$\xi = \pm\frac{1}{(2\pi)^{3/2}}\sqrt{\frac{\hbar}{8\alpha\varepsilon_0 c^3}} = \frac{\hbar}{4\pi e c} = \pm 1.747\ 10^{-25}\ Volt\ m^{-1}s^2 \quad (51)$$

where $\alpha = e^2/4\pi\varepsilon_0\hbar c \approx 1/137.036$ is the fine structure constant and e is the electron–positron charge.

Obviously, when introducing Eq. (50) in Eq. (17) for a single k-mode photon, an appropriate volume V_k has to be considered for the equation to hold:

$$E_k = \hbar\omega_k = 2\varepsilon_0 V_k \xi^2 \omega_k^4 \quad (52)$$

Thus, the characteristic volume of a free single photon is written, in agreement with Eq. (47), as

$$V_k = \left(\frac{\hbar}{2\varepsilon_0\xi^2}\right)\omega_k^{-3} \quad (53)$$

Replacing ξ expressed by Eq. (51) in Eq. (53), we obtain for the photon quantization volume

$$V_k \simeq 4\alpha\lambda_k^3 \tag{54}$$

Equations (50) and (53) express the quantized vector potential amplitude and the spatial extension of a single photon with the constant ξ evaluated to be $\xi = \hbar/4\pi ec$.

4.2. *Photon classical–quantum (wave–particle) physical properties*

For a free k-mode photon, the volume V_k corresponds to the space in which the quantized vector potential oscillates at the angular frequency ω_k over a period along the propagation axis, generating orthogonal electric and magnetic fields, whose amplitudes are, respectively,

$$|\vec{\varepsilon}_k| = |-\partial\vec{\alpha}_{k\lambda}(\vec{r},t)/\partial t| \propto |\xi|\omega_k^2; \quad |\vec{\beta}_k| \propto \sqrt{\varepsilon_0\mu_0}|\xi|\omega_k^2 \tag{55}$$

which are independent of any external arbitrary volume parameter and are directly proportional to the square of the angular frequency [55, 56].

We can now express the quantum properties of the photon, energy, momentum and spin by integrating the classical electromagnetic expressions over the volume V_k and by using the vector potential amplitude obtained in Eq. (50), linking by this way the classical (wave) to the quantum (particle) representations [2]. The energy is written as

$$E_k = \int_{V_k} 2\varepsilon_0\alpha_{0k}^2\omega_k^2 d^3r = \int_{V_k} 2\varepsilon_0\xi^2\omega_k^4 d^3r = 2\varepsilon_0\xi^2\omega_k^4 V_k = \hbar\omega_k \tag{56}$$

With the same token, considering circular polarization [4, 5, 7, 9] for the amplitudes of the electric and magnetic fields in Eq. (55), the momentum is

$$\begin{aligned}\vec{p}_k &= \int_{V_k} \varepsilon_0\vec{\varepsilon}_{k\lambda} \times \vec{\beta}_{k\lambda} d^3r \\ &= \varepsilon_0\left(\sqrt{2}\omega_k\alpha_{0k}\right)\left(\sqrt{2}\omega_k\alpha_{0k}/c\right) V_k\vec{k}/|\vec{k}| = \hbar\vec{k}\end{aligned} \tag{57}$$

According to the classical electromagnetic theory, the spin can be written through the electric and magnetic field components, hence, using again the circular polarization, we get

$$|\vec{S}| = \int_{V_k} \varepsilon_0 |\vec{r} \times (\vec{\varepsilon}_{k\lambda} \times \vec{\beta}_{k\lambda})| d^3r$$
$$= \varepsilon_0 (c/\omega_k) \left(\sqrt{2}\omega_k \alpha_{0k}\right) \left(\sqrt{2}\omega_k \alpha_{0k}/c\right) V_k = \hbar \qquad (58)$$

where we have taken the mean value $\langle|\vec{r}|\rangle_k = c/\omega_k$ obtained for a single photon state [57].

The fact that the quantum properties, energy, momentum and spin of the photon can be expressed through the classical electromagnetic fields integrated over the volume V_k signifies that the photon has naturally a spatial extension and consequently when employing the term "wave–particle," one must have in mind that a single photon is a "three-dimensional entity".

We can now obtain Heisenberg's uncertainty relation for position and momentum using V_k. Indeed, replacing V in Eq. (16) by V_k, we get the photon vector potential amplitude operators

$$\tilde{\alpha}_{0k\lambda} = \xi\omega_k \alpha_{k\lambda}; \quad \tilde{\alpha}^*_{0k\lambda} = \xi\omega_k \alpha^+_{k\lambda} \qquad (59)$$

The corresponding position $\tilde{Q}_{k\lambda}$ and momentum $\tilde{P}_{k\lambda}$ Hermitian operators [19, 30, 52] are written as

$$\tilde{Q}_{k\lambda} = \sqrt{\varepsilon_0 V_k}\,(\tilde{\alpha}_{0k\lambda} + \tilde{\alpha}^*_{0k\lambda})\,; \quad \tilde{P}_{k\lambda} = -i\omega_k \sqrt{\varepsilon_0 V_k}\,(\tilde{\alpha}_{0k\lambda} - \tilde{\alpha}^*_{0k\lambda}) \qquad (60)$$

Thus, introducing Eq. (59) in Eq. (60) and using Eq. (20) with Eq. (53), Heisenberg's commutation relation, a fundamental concept in quantum theory, results directly [2]

$$[\tilde{Q}_{k\lambda}, \tilde{P}_{k'\lambda'}] = -i\varepsilon_0 \omega^2_{k'} \omega_k \sqrt{V_k V_{k'}}\left[\left(\xi a_{k\lambda} + \xi a^+_{k\lambda}\right), \left(\xi a_{k'\lambda'} - \xi a^+_{k'\lambda'}\right)\right]$$
$$= i\hbar \delta_{kk'} \delta_{\lambda\lambda'} \qquad (61)$$

The fundamental properties of the photon, energy E_k, momentum $\vec{p}_k$ and wave vector $\vec{k}$, are complemented by the vector potential amplitude α_{0k}, expressing its electromagnetic nature

$$E_k/\hbar = |\vec{p}_k| c/\hbar = |\vec{k}| c = \alpha_{0k}/|\xi| = \omega_k \qquad (62)$$

Considering Heisenberg's *energy–time* uncertainty principle

$$\delta E_k\, \delta t \geq \hbar \tag{63}$$

we directly deduce from Eq. (62) the *vector potential–time* uncertainty

$$\delta \alpha_{0k} \delta t \geq |\xi| \tag{64}$$

The energy and vector potential uncertainty with respect to time are intrinsic physical properties of the wave–particle nature of the photon.

4.3. *Photon wave–particle equation and wave function*

Obviously, the photon vector potential function $\vec{\alpha}_{k\lambda}(\vec{r}, t)$ expressed in Eq. (49) satisfies the wave propagation equation in vacuum issued from Maxwell's equations

$$\vec{\nabla}^2 \vec{\alpha}_{k\lambda}(\vec{r}, t) - \frac{1}{c^2} \frac{\partial^2}{\partial t^2} \vec{\alpha}_{k\lambda}(\vec{r}, t) = 0 \tag{65}$$

as well as the *vector potential–energy* (*wave–particle*) equation for the photon, [2, 55]

$$i \begin{pmatrix} \xi \\ \hbar \end{pmatrix} \frac{\partial}{\partial t} \vec{\alpha}_{k\lambda}(\vec{r}, t) = \begin{pmatrix} \tilde{\alpha}_{0k} \\ \tilde{H} \end{pmatrix} \vec{\alpha}_{k\lambda}(\vec{r}, t) \tag{66}$$

where the vector potential operator $\tilde{\alpha}_{0k} = -i\xi c\vec{\nabla}$ and the relativistic Hamiltonian for a massless particle $\tilde{H} = -i\,\hbar\, c\vec{\nabla}$ have the eigenvalues $\xi\omega_k$ and $\hbar\omega_k$, respectively [2, 54].

It is worthwhile to note the symmetry between the pairs $\{E_k, \hbar\}$ and $\{\alpha_{0k}, \xi\}$ for a single photon characterizing, respectively, the particle (energy) and electromagnetic wave (vector potential) natures, bearing in mind that the energy corresponds to the integration of the single-mode electromagnetic field energy density over the volume V_k.

Now, when considering the propagation of a k-mode photon with wavelength λ_k, the difficulties for defining a position operator are widely discussed in the literature [28, 30, 31, 40–42, 58]. At the same time, the efforts for defining a wave function for the photon based on the electric and magnetic fields were rather fruitless [59–63]. It has been emphasized several times [19,27,28,31,58] that a photon cannot be localized along the propagation axis in a length shorter than the wavelength λ_k and within a volume smaller than roughly λ_k^3.

In fact, from a theoretical point of view, a photon propagating in the z-direction Heisenberg uncertainty for the position z and momentum $P_z = \hbar k_z = h/\lambda_k$ is written as

$$\delta z\, \delta P_z \geq h \rightarrow \delta z\, \delta(1/\lambda_k) \geq 1 \tag{67}$$

Note that the momentum uncertainty along the propagation axis is expressed through the uncertainty over the inverse of the wavelength.

Considering now the vector potential function with the quantized amplitude $\alpha_{0k} = \xi\omega_k$ as a real wave function for the photon, when a k-mode photon is emitted at a coordinate z_e at an instant t_e and propagates in vacuum along the z-axis, the probability $\Pi_k(z)$ to be localized at time t and at the coordinate $z = z_e + c(t - t_e)$ corresponds to the square of the modulus of the vector potential and is consequently proportional to the square of the angular frequency [55, 56]:

$$\Pi_k(z) \propto |\vec{\alpha}_{k\lambda}(z,t)|^2 = \xi^2\omega_k^2 \propto \lambda_k^{-2} \tag{68}$$

Obviously, the shorter the wavelength of the photon, the higher the localization probability, in agreement with Heisenberg's uncertainty and the experimental evidence.

4.4. *Electromagnetic vacuum, photons, electrons–positrons*

The photon vector potential is composed of a fundamental function $\Xi_{k\lambda}$ times the angular frequency ω_k and is written in the classical (wave) and quantum (particle) formalisms as

$$\vec{\alpha}_{k\lambda} = \xi\,\omega_k \left[\hat{\varepsilon}_{k\lambda} e^{i\left(\vec{k}\cdot\vec{r} - \omega_k t + \phi\right)} + \hat{\varepsilon}^*_{k\lambda} e^{-i\left(\vec{k}\cdot\vec{r} - \omega_k t + \phi\right)}\right] = \omega_k \vec{\Xi}_{k\lambda}(\omega_k, \vec{r}, t) \tag{69}$$

$$\begin{aligned}\tilde{\alpha}_{k\lambda} &= \xi\,\omega_k \left[a_{k\lambda}\hat{\varepsilon}_{k\lambda} e^{i\left(\vec{k}\cdot\vec{r} - \omega_k t + \phi\right)} + a^+_{k\lambda}\hat{\varepsilon}^*_{k\lambda} e^{-i\left(\vec{k}\cdot\vec{r} - \omega_k t + \phi\right)}\right] \\ &= \omega_k \tilde{\Xi}_{k\lambda}(a_{k\lambda}, a^+_{k\lambda})\end{aligned} \tag{70}$$

In this way, the general equation for the vector potential of the electromagnetic wave considered as a superposition of plane wave modes is written as

$$\begin{aligned}\vec{A}(\vec{r},t) &= \sum_{k,\lambda} \xi\, \omega_k \left[\hat{\varepsilon}_{k\lambda} e^{i\left(\vec{k}\cdot\vec{r}-\omega_k t+\phi\right)} + \hat{\varepsilon}^*_{k\lambda} e^{-i\left(\vec{k}\cdot\vec{r}-\omega_k t+\phi\right)}\right] \\ &= \sum_{k,\lambda} \omega_k \vec{\Xi}_{k\lambda}(\omega_k, \vec{r}, t) \end{aligned} \tag{71}$$

and that of a large number of cavity-free photons in QED is

$$\begin{aligned}\tilde{A} &= \sum_{k,\lambda} \xi\, \omega_k \left[a_{k\lambda}\hat{\varepsilon}_{k\lambda} e^{i\left(\vec{k}\cdot\vec{r}-\omega_k t+\phi\right)} + a^+_{k\lambda}\hat{\varepsilon}^*_{k\lambda} e^{-i\left(\vec{k}\cdot\vec{r}-\omega_k t+\phi\right)}\right] \\ &= \sum_{k,\lambda} \omega_k \tilde{\Xi}_{k\lambda}(a_{k\lambda}, a^+_{k\lambda}) \end{aligned} \tag{72}$$

According to Eqs. (55) and (62), for $\omega_k \to 0$, all the physical properties of the photon vanish, entailing that the photon exists only for a non-zero frequency of the vector potential oscillation. However, the zero-frequency level does not correspond to perfect inexistence because the fundamental field $\Xi_{k\lambda}$ does not vanish for $\omega_k = 0$ but reduces to $\Xi_{0\lambda}$ involving the amplitude ξ and the general expression of the polarization vectors $\widehat{\varepsilon}_{k\lambda}$ [64, 65], and is written in the classical and quantum representations as

$$\vec{\Xi}_{0\lambda} = \xi \left[\hat{\varepsilon}_{k\lambda} e^{i\phi} + \hat{\varepsilon}^*_{k\lambda} e^{-i\phi}\right]; \quad \tilde{\Xi}_{0\lambda} = \xi \left[a_{k\lambda}\hat{\varepsilon}_{k\lambda} e^{i\phi} + a^+_{k\lambda}\hat{\varepsilon}^*_{k\lambda} e^{-i\phi}\right] \tag{73}$$

The field $\Xi_{0\lambda}$ is the electromagnetic field ground state permeating all the space ($\lambda_k \to \infty$) and having zero-energy and zero-vector potential as well as zero electric and magnetic fields. This physical state lies beyond the Bohm–Aharonov situation in which the energy, the electric and magnetic fields are zero, but a vector potential is present in space [44]. Thus, in the complete absence of energy and vector potential, the field $\Xi_{0\lambda}$ corresponds to the electromagnetic vacuum whose states constitute the main "skeleton" of photons, which now clearly appear to be a vacuum oscillation [2, 64, 65].

Combination of the expression $\xi = \hbar/4\pi ec$ with the fine structure constant definition $\alpha = e^2/4\pi\varepsilon_0\hbar c$ permits us to deduce directly the electron–positron elementary charge, $e = \pm 1.602\ 10^{-19}C$, a fundamental physical constant, which is now expressed exactly through the electromagnetic vacuum amplitude ξ [65, 66]:

$$e = (4\pi)^2 \alpha \frac{\xi}{\mu_0} \tag{74}$$

Using again Eq. (51) and recalling that the electron mass may be written as $m_e = e\hbar/2\mu_B$, where μ_B is the Bohr magneton, we deduce that the electron mass is also expressed as a function of the amplitude ξ [65]:

$$m_{e^-,e^+} = 2\pi c\, e^2 \frac{\xi}{\mu_B} \tag{75}$$

entailing that the mass also issues from the electromagnetic vacuum.

Equations (50), (74) and (75) show the strong physical relationship between photons and electrons–positrons which are all related directly to the electromagnetic vacuum through the amplitude ξ. Obviously, photons and electrons–positrons, also probably leptons–antileptons, are issued from the same quantum vacuum field. This may be at the origin of the physical mechanism governing the photon generation during the electron–positron (and probably lepton–antilepton) annihilation and that of the electron–positron (lepton–antilepton) pair creation during the annihilation of high-energy gamma photons in the vicinity of a very heavy nucleus.

5. Conclusion

In this chapter, we have presented the recent theoretical developments, complementing the standard QED formalism with the purpose of describing a single photon state in conformity with the experiments. We summarize the principal features in the following.

The quantization of the vector potential amplitude α_{0k}, a real physical entity, for a single free-of-cavity k-mode photon with angular frequency ω_k is expressed by $\alpha_{0k} = \xi\omega_k$, where $\xi = \hbar/4\pi ec$, and leads to the establishment of a *vector potential–energy* (electromagnetic wave–particle) formalism for the photon (66), expressing the the first (energy) and second (vector potential) quantization of the electromagnetic field. A single photon state is an indivisible entity of the electromagnetic field, extending over a wavelength λ_k and consisting of the quantized vector potential oscillating at the angular frequency ω_k, with circular polarization, giving birth to orthogonally oscillating electric and magnetic fields whose amplitudes are proportional to the square of the angular frequency $\xi\omega_k^2$ (55). Its lateral expansion, confirmed experimentally, yields a minimum photon volume V_k, which is proportional to λ_k^3 in agreement with the experimental observations. The quantum properties of the photon, energy, momentum and spin are obtained directly from the classical electromagnetic expressions integrated over the volume V_k (see Eqs. (56)–(58)). It is also shown (see Eq. (61)) that the Heisenberg uncertainty can be readily obtained through the volume V_k.

A single photon, as a local three-dimensional entity of the electromagnetic field, propagates guided by the non-local vector potential function (49), which appears to be a natural wave function for the photon, satisfying both the propagation equation (65) and the *vector potential–energy* equation (66). The probability for detecting a photon around a given point on the propagation axis is obtained by the square modulus of the vector potential and is proportional to the square of the angular frequency $\xi^2\omega_k^2$ (see Eq. (68)) which signifies that the higher the frequency, the better the localization, in agreement with the experiments.

Finally, the electromagnetic vacuum issues naturally from the vector potential wave function, putting in evidence that photons are oscillations of the vacuum field. Furthermore, the electron–positron charge and mass are directly proportional to the vector potential amplitude quantization constant, showing the strong physical relationship with the photons. Obviously, the origin of the mechanisms governing the transformations between photons and electrons–positrons lies in the nature of the electromagnetic vacuum.

References

[1] V. Ronchi, *The Nature of Light: An Historical Survey* (Harvard University Press, Cambridge, MA, USA, 1970).

[2] C. Meis, *Light and Vacuum*, 2nd ed. (World Scientific, Singapore, 2017).

[3] I. Newton, *Opticks* (Dover, New York, NY, USA, 1952).

[4] B. E. A. Saleh and M. C. Teich, *Fundamentals of Photonics* (John Wiley & Sons, Hoboken, NJ, USA, 2007).
[5] F. H. Read, *Electromagnetic Radiation* (John Wiley & Sons, New York, NY, USA, 1980).
[6] M. Born and E. Wolf, *Principles of Optics* (Cambridge University Press, Cambridge, UK, 1999).
[7] J. D. Jackson, *Classical Electrodynamics* (Wiley, New York, NY, USA, 1998).
[8] M. Planck, *The Theory of Heat Radiation* (Dover, New York, NY, USA, 1959).
[9] B. H. Bransden and C. J. Joachain, *Physics of Atoms and Molecules*, 2nd ed. (Pearson, London, UK, 2003).
[10] A. A. Michelson *et al.*, Conference on the Michelson-Morley experiment, *Astrophys. J.* 68, 341 (1928).
[11] H. Hertz, Ueber einen Einfluss des ultravioletten Lichtes auf die electrische Entladung, *Ann. Phys.* 267, 983 (1887).
[12] A. Einstein, *The Collected Papers of Albert Einstein*, J. Stachel, D. C. Cassidy, J. Renn, and R. Schulmann, Eds. (Princeton University Press, Princeton, NJ, USA, 1987).
[13] N. Bohr, H. A. Kramers, and J. C. Slater, The quantum theory of radiation, *Philos. Mag.* 47, 785 (1924).
[14] G. I. Taylor, Interference fringes with feeble light, *Proc. Camb. Philos. Soc.* 15, 114 (1909).
[15] A. H. Compton, A quantum theory of the scattering of X-rays by light elements, *Phys. Rev.* 21, 483 (1923).
[16] G. Wentzel, Zur Theorie des photoelektrischen Effekts, *Z. Phys.* 40, 574 (1926).
[17] G. Beck, Zur Theorie des Photoeffekts, *Z. Phys.* 41, 443 (1927).
[18] W. E. Lamb, Jr. and M. O. Scully, in *Polarisation, Matière et Rayonnement, Volume Jubilaire en l'Honneur d'Alfred Kastler* (French Physical Society, Presses Universitaires de France, Paris, France, 1969).
[19] J. C. Garrison and R. Y. Chiao, *Quantum Optics* (Oxford University Press, Oxford, UK, 2008).
[20] O. Klein and Y. Nishina, Über die Streuung von Strahlung durch freie Elektronen nach der neuen relativistischen Quantendynamik von Dirac, *Z. Phys.* 52, 853 (1929).
[21] F. Jin, S. Yuan, H. De Raedt, K. Michielsen, and S. Miyashita, Corpuscular model of two-beam interference and double-slit experiments with single photons, *J. Phys. Soc. Jpn.* 79, 074401 (2010).
[22] R. P. Feynman, *QED: The Strange Theory of Light and Matter* (Princeton University Press, Princeton, NJ, USA, 1988).

[23] A. L. Schawlow and C. H. Townes, Infrared and optical masers, *Phys. Rev.* 112, 1940 (1958).
[24] D. N. Klyshko, A. N. Penin, and B. F. Polkovnikov, Parametric luminescence and light scattering by polaritons, *JETP Lett.* 11, 5 (1970).
[25] D. C. Burnham and D. L. Weinberg, Observation of simultaneity in parametric production of optical photon, *Phys. Rev. Lett.* 25, 84 (1970).
[26] P. Ghose and D. Home, The two-prism experiment and wave-particle duality of light, *Found. Phys.* 26, 943 (1996).
[27] A. I. Akhiezer and V. B. Berestetskii, *Quantum Electrodynamics* (Interscience, New York, NY, USA, 1965).
[28] L. Mandel, Configuration-space photon number operators in quantum optics, *Phys. Rev.* 144, 1071 (1966).
[29] R. L. Pfleegor and L. Mandel, Interference of independent photon beams, *Phys. Rev.* 159, 1084 (1967).
[30] G. Grynberg, A. Aspect, and C. Fabre, *Introduction to Quantum Optics* (Cambridge University Press, Cambridge, UK, 2010).
[31] G. Auletta, *Foundations and Interpretation of Quantum Mechanics* (World Scientific, Singapore, 2001).
[32] P. Grangier, G. Roger, and A. Aspect, Experimental evidence for a photon anticorrelation effect on a beam splitter: A new light on single-photon interferences, *Europhys. Lett.* 1, 173 (1986).
[33] H. L. Robinson, Diffraction patterns in circular apertures less than one wavelength in diameter, *J. Appl. Phys.* 24, 35 (1953).
[34] R. K. Hadlock, Diffraction patterns at the plane of a slit in a reflecting screen, *J. Appl. Phys.* 29, 918 (1958).
[35] G. Hunter and R. L. P. Wadlinger, Photons and neutrinos as electromagnetic solitons, *Phys. Essays* 2, 158 (1989).
[36] G. Hunter, in *Quantum Uncertainties: Recent and Future Experiments and Interpretations*, W. M. Honig, D. W. Kraft, and E. Panarella, Eds., NATO ASI Series (Series B: Physics), Vol. 162 (Springer, Boston, MA, USA, 1987), p. 331.
[37] W. Heitler, *The Quantum Theory of Radiation* (Clarendon Press, Oxford, UK, 1954).
[38] R. P. Feynman, *QED: The Strange Theory of Light and Matter* (Princeton University Press, Princeton, NJ, USA, 1988).
[39] L. H. Ryder, *Quantum Field Theory* (Cambridge University Press, Cambridge, UK, 1987).
[40] P. W. Milonni, *The Quantum Vacuum: An Introduction to Quantum Electrodynamics* (Academic Press, San Diego, CA, USA, 1994).
[41] M. H. Mittleman, *Introduction to the Theory of Laser-Atom Interactions* (Plenum Press, New York, NY, USA, 1982).

[42] M. Weissbluth, *Photon-Atom Interactions* (Academic Press, Boston, MA, USA, 1988).
[43] W. Ehrenberg and R. E. Siday, The refractive index in electron optics and the principles of dynamics, *Proc. Phys. Soc. London, Sect. B* 62, 8 (1949).
[44] Y. Aharonov and D. Bohm, Significance of electromagnetic potentials in the quantum theory, *Phys. Rev.* 115, 485 (1959).
[45] R. G. Chambers, Shift of an electron interference pattern by enclosed magnetic flux, *Phys. Rev. Lett.* 5, 3 (1960).
[46] A. Tonomura *et al.*, Observation of the Aharonov-Bohm effect by electron holography, *Phys. Rev. Lett.* 48, 1443 (1982).
[47] N. Osakabe *et al.*, Experimental confirmation of Aharonov-Bohm effect using a toroidal magnetic field confined by a superconductor, *Phys. Rev. A* 34, 815 (1986).
[48] T. A. Welton, Some observable effects of the quantum-mechanical fluctuations of the electromagnetic field, *Phys. Rev.* 74, 1157 (1948).
[49] P. W. Milonni, Casimir forces without the vacuum radiation field, *Phys. Rev. A* 25, 1315 (1982).
[50] J. Schwinger, L. L. DeRaad, Jr., and K. A. Milton, Casimir effect in dielectrics, *Ann. Phys.* 115, 1 (1978).
[51] H. Ezawa, K. Nakamura, and K. Watanabe, The Casimir force from Lorentz's approach, in *Frontiers in Quantum Physics: Proceedings of the 2nd International Conference, Kuala Lumpur, Malaysia, 9–11 July 1997*, S. C. Lim, R. Abd-Shukor, and K. H. Kwek, Eds. (Springer, Singapore, 1998), pp. 160–169.
[52] C. Cohen-Tannoudji, B. Diu, and F. Laloë, *Mécanique Quantique*, Tome III (EDP Sciences, Les Ulis, France, 2017).
[53] H. Walther, B. T. H. Varcoe, B.-G. Englert, and T. Becker, Cavity quantum electrodynamics, *Rep. Prog. Phys.* 69, 1325 (2006).
[54] C. Meis, Photon wave-particle duality and virtual electromagnetic waves, *Found. Phys.* 27, 865 (1997).
[55] C. Meis and P. R. Dahoo, Vector potential quantization and the photon intrinsic electromagnetic properties: Towards nondestructive photon detection, *Int. J. Quantum Inf.* 15, 1740003 (2017).
[56] C. Meis and P. R. Dahoo, Quantized vector potential and the photon wave-function, *J. Phys.: Conf. Ser.* 936, 012004 (2017).
[57] Yu. P. Rybakov and B. Saha, Soliton model of atom, *Found. Phys.* 25, 1723 (1995).
[58] S. Jeffers, S. Roy, J.-P. Vigier, and G. Hunter, Eds., *The Present Status of the Quantum Theory of Light* (Kluwer Academic, Dordrecht, The Netherlands, 1997).

[59] I. Bialynicki-Birula, Photon wave function, in *Progress in Optics*, Vol. XXXVI, E. Wolf, Ed. (Elsevier, Amsterdam, The Netherlands, 1996).

[60] N. Chandrasekar, Quantum mechanics of photons, *Adv. Stud. Theor. Phys.* 6, 1 (2012).

[61] J. E. Sipe, Photon wave functions, *Phys. Rev. A* 52, 1875 (1995).

[62] B. J. Smith and M. G. Raymer, Photon wave functions, wave-packet quantization of light, and coherence theory, *New J. Phys.* 9, 414 (2007).

[63] D. L. Khokhlov, Spatial and temporal wave functions of photons, *Appl. Phys. Res.* 2, 1 (2010).

[64] C. Meis, Vector potential quantization and the quantum vacuum, *Phys. Res. Int.* 2014, 187432 (2014).

[65] C. Meis, The electromagnetic field ground state and the cosmological evolution, *J. Phys.: Conf. Ser.* 1141, 012072 (2018).

[66] C. Meis and P. R. Dahoo, The single photon state, the quantum vacuum and the elementary electron-positron charge, *AIP Conf. Proc.* 2040, 020011 (2018).

Chapter 6

Photon Structure and Wave Function from the Vector Potential Quantization*

Abstract

A photon structure is advanced based on the experimental evidence and the vector potential quantization at a single photon level. It is drawn that the photon is neither a point particle nor an infinite wave but behaves rather like a local "wave–corpuscle" extended over a wavelength, occupying a minimum quantization volume and guided by a non-local vector potential real wave function. The quantized vector potential oscillates over a wavelength with circular left or right polarization, giving birth to orthogonal magnetic and electric fields whose amplitudes are proportional to the square of the frequency. The energy $\hbar\omega$ and momentum $\hbar\vec{k}$ are carried by the local wave–corpuscle propagating guided by the non-local vector potential wave function, which is suitably normalized.

1. Introduction

A single photon is a particular relativistic massless wave–particle, for which scientists still face major difficulties defining a clear physical representation. Historically, the scientific understanding of light's

*This chapter is reproduced from C Meis. 2023. "Photon Structure and Wave Function from the Vector Potential Quantization", *Journal of Modern Physics*, 14(3), pp. 311–329.

properties, guided each time by the interpretation of the experiments, was a continuous balancing between the wave and particle natures. Before going ahead to the quantum description of the photons and the possible associated wave functions, it is of crucial importance to understand the evolution of our concepts about the nature of light.

In the seventeenth century, reviving the ideas of ancient Greeks on light's nature, Newton advanced that light is composed of individual corpuscles which travel rectilinearly in a homogeneous medium [1,2]. On the other side, Huygens developed a refined theory of light [3,4] based on the wave representation and refuted Newton's corpuscular theory. Almost two centuries later, Young experimentally obtained interference patterns using different light sources. He also went further by explaining some polarization experiments, assuming that light oscillations take particular orientations with respect to the propagation axis [4, 5]. By that time, it was impossible to explain the experimental diffraction patterns using Newton's corpuscular theory, while Euler and Fresnel advanced easily precise interpretations, applying the wave representation. James Clerk Maxwell published his remarkable theory on electromagnetic waves in 1865. He established, for the first time, the relations between the electric and magnetic fields and advanced that light is composed of electromagnetic waves [6]. Hertz confirmed this experimentally by discovering the long-wavelength electromagnetic radiation. Thus, by the end of the nineteenth century, the scientific community had replaced Newton's corpuscular representation with the wave theory of light. However, that was not for a long time.

At the beginning of the twentieth century, in order to explain the spectral energy density emitted by a black body, Max Planck reintroduced the notion of the particle nature of light in a particular sense. In fact, he assumed that bodies are composed of "oscillators" having the particularity to emit electromagnetic "packets," each with energy $h\nu$, where ν is the frequency and h is Planck's constant [7,8]. A few years later, based on Planck's works, Einstein proposed an interpretation of the photoelectric effect, first observed by Hertz [9], assuming that the electromagnetic radiation itself is composed of quanta with energy $h\nu$ [10]. Furthermore, in 1923, Compton advanced that only the light quanta could explain the experimental observations of the X-rays scattering on free electrons [11]. Hence, the photoelectric effect and Compton X-ray scattering have always been considered as the strongest arguments in favor of the particle nature of light, and the quantum electrodynamics (QED) theory was developed from the 1930s to the 1960s based on the point photon model.

However, it is scarcely quoted in the literature that some important studies in favor of the electromagnetic wave theory have been totally disregarded. In fact, Wentzel in 1926 [12], Beck in 1927 [13], as well as Lamb and Scully in the 1960s [14] demonstrated that the photoelectric effect can be interpreted directly by considering only the electromagnetic wave nature of light and without referring to photons at all. Furthermore, Klein and Nishina in 1929 [15] fully interpreted Compton's scattering by also considering the wave nature of light and without invoking the photon concept. So, even after the Second World War, although the majority of scientists joined Bohr's *complementarity principle*, according to which *light exhibits both wave and particle natures appearing mutually exclusively in each experiment*, a significant part of the scientific community still had the conviction that the electromagnetic wave representation was sufficient to understand light's nature.

Things radically changed in the fifties and the sixties. Robinson in 1953 [16] and Hadlock in 1958 [17] carried out systematic experiments using microwaves, crossing variable dimensions rectangular or circular apertures and deduced that no energy is transmitted when the aperture dimensions are smaller than approximately a quarter of the wavelength ($\sim\lambda/4$). This was also confirmed later by Hunter and Wadlinger [18, 19] using X-band microwaves. On the other hand, Mandel's experiments in the sixties [20, 21], employing the recently discovered laser technics, concluded that a single photon has circular polarization and cannot be localized in a distance shorter than its wavelength λ [22], and more generally, it cannot be better localized than within a volume of the order of the cube of its wavelength (λ^3) [23, 24]. Consequently, the experimental evidence conflicts with both the point photon model, upon which QED has been developed, and the classical continuous electromagnetic wave theory, showing that light must be composed of localized "wave packets" traveling in vacuum at the universal velocity c [25] and carrying energy $h\nu$ and momentum $h\nu/c$. In reality, the point photon concept has permitted us to establish an extremely efficient mathematical approach in QED for describing states before and after an interaction process [26–28].

However, it is obvious that it is inappropriate for the description of the real nature of a single photon state.

The development of the revolutionary parametric down-conversion techniques [29, 30] in the seventies permitted us to realize the conditions in which, with an excellent statistical confidence, only a single photon is present in the experimental device. Employing these techniques, the double prism experiment [31] carried out in the nineties contradicted Bohr's *mutual exclusiveness principle,* showing that a *single photon exhibits both the wave and particle natures in the same experiment.* Furthermore, Grangier *et al.* demonstrated experimentally the indivisibility of photons [28, 32], while the entangled states, first observed by Kosher and Commins [33] and further investigated in the eighties [34, 35], has shown that the photon should be locally an integral entity during the detection procedure but with a real non-local wave function.

Finally, a synthesis of the experimental studies shows that a single k-mode photon is a local, indivisible segment of the electromagnetic field with circular left or right polarization corresponding to spin $\pm h/2\pi$, respectively. It carries a quantum of energy $h\nu$ and has a momentum $h\nu/c$. Its intrinsic spatial length extends along a wavelength λ_k and can only be detected within a volume $\sim\lambda_k^3$, yielding that its radial expansion should be proportional to a fraction of its wavelength. Consequently, it seems to be a local "wave–corpuscle" absorbed and emitted as a whole and guided by a non-local wave function.

All these physical characteristics yield particular difficulties for the description of a real single photon state and consequently for the definition of an appropriate wave function within the quantum mechanics concepts.

In what follows, we take into account all the above experimental and theoretical facts in order to establish a physical picture of the single photon and its wave function. We first give a brief presentation of the link between the classical and quantum theories through the vector potential, and then we analyze the second quantization process, resulting in a quantum description of the electromagnetic field. Then, we address the spatial properties of a single photon state and enhance the quantization of the vector potential amplitude to a single photon level, getting a physical representation conform to the experimental evidence. Next, we advance a photon wave function based on the vector potential quantization, satisfying the wave propagation equation, Schrodinger's equation with the relativistic massless particle Hamiltonian, an equivalent equation for the vector potential, and finally, we discuss the characteristic properties of the established photon wave function and its normalization.

2. The electromagnetic field vector potential in classical and quantum theories

2.1. *The vector potential: Classical to quantum link*

Experiments have shown that the vector potential is not a mathematical artifact but a real physical field exerting a direct influence on charges [36–39]. As that, it represents the fundamental link between the classical electromagnetic wave theory issued from Maxwell's equations and QED [8, 22, 40, 41]. In the classical theory [3–6], the energy density of a mode k of the electromagnetic wave depends on the square of the modulus of the electric field $\vec{E}_k(\vec{r},t)$ and the magnetic field $\vec{B}_k(\vec{r},t)$, and is written as

$$W_k(\vec{r},t) = \frac{1}{2}\left(\varepsilon_0|\vec{E}_k(\vec{r},t)|^2 + \frac{1}{\mu_0}|\vec{B}_k(\vec{r},t)|^2\right) \tag{1}$$

with ε_0 and μ_0 being the electric permittivity and magnetic permeability of the vacuum, respectively, satisfying the relation $\varepsilon_0\mu_0 c^2 = 1$, where c is the speed of light in vacuum.

For a monochromatic plane wave mode k with angular frequency ω_k and vector potential amplitude $a_{0k}(\omega_k)$, we have

$$\vec{E}_k(\vec{r},t) = -2\omega_k a_{0k}\,(\omega_k)\,\hat{\varepsilon}\sin(\vec{k}\cdot\vec{r}-\omega_k t) \tag{2}$$

$$\vec{B}_k(\vec{r},t) = -\frac{1}{c}2\omega_k a_{0k}\,(\omega_k)\,(\hat{k}\times\hat{\varepsilon})\sin(\vec{k}\cdot\vec{r}-\omega_k t) \tag{3}$$

with $\hat{\varepsilon}$ the unit vector perpendicular to the propagation axis, $|\vec{k}| = 2\pi/\lambda_k$ the wave vector along the propagation axis and λ_k the wavelength of the mode k.

Using (2) and (3) in Eq. (1), the energy density is written as a function depending uniquely on the squares of the vector potential amplitude and of the angular frequency

$$W_k(\vec{r},t) = 4\varepsilon_0\omega_k^2|a_{0k}(\omega_k)|^2 \sin^2(\vec{k}\cdot\vec{r} - \omega_k t) \tag{4}$$

The mean value of the last expression over a period, i.e., over a wavelength, is

$$\langle W_k \rangle = 2\varepsilon_0\omega_k^2 \, |a_{0k}(\omega_k)|^2 \tag{5}$$

It is important to note that the mean energy density $\langle W_k \rangle$ is independent of any external volume parameter. Hence, in the classical theory, a free-of-cavity electromagnetic radiation mode occupies naturally a minimum volume, while in a resonant cavity, this volume corresponds roughly to that imposed by the boundary conditions and the cut-off wave vectors for the mode [4, 42].

Now, in the quantum description, the energy density in a given volume V for a number $n(\omega_k)$ of k-mode photons, each with angular frequency ω_k and energy $\hbar\omega_k$ is simply

$$W_k = \frac{n(\omega_k)\hbar\omega_k}{V} \tag{6}$$

The link between the classical and quantum theories is established by imposing the classical mean energy density over a period (5) to be equal to that of the quantum description (6). Thus, considering $n(\omega_k) = 1$, we get the vector potential amplitude for a single k-mode photon:

$$a_{0k}(\omega_k) = \sqrt{\frac{\hbar}{2\varepsilon_0\omega_k V}} \tag{7}$$

The last relation constitutes the fundamental link between the classical and quantum theories of light, and it is applied in QED in order to define the vector potential amplitude operators for a single photon [23, 27, 28]:

$$\tilde{a}_{k\lambda} = \sqrt{\frac{\hbar}{2\varepsilon_0\omega_k V}} a_{k\lambda} \quad \tilde{a}^*_{k\lambda} = \sqrt{\frac{\hbar}{2\varepsilon_0\omega_k V}} a^+_{k\lambda} \tag{8}$$

where $a_{k\lambda}$ and $a^+_{k\lambda}$ are, respectively, the annihilation and creation non-Hermitian operators for a single k-mode and λ-polarization photon.

Obviously, this procedure introduces an external volume parameter V in the definitions for the vector potential amplitude operators of the single photon. Consequently, one could draw out that single photons in free space, in other words, in a cavity with infinite dimensions, should have zero vector potential amplitudes and thus zero energy. This ambiguity, which is scarcely quoted in the literature, automatically implies that for a single photon, the volume V in (8) cannot be an arbitrary external parameter but corresponds roughly to that defined by the boundary conditions in a cavity for the single radiation mode k.

2.2. *The electromagnetic field vector potential in QED*

In quantum field theory [23, 28, 40], the radiation vector potential is written as

$$\vec{A}(\vec{r},t) = \int \frac{d^3k}{(2\pi)^{3/2}} \sqrt{\frac{\hbar}{2\varepsilon_0 \omega_k}} \sum_{\lambda=1}^{2} \left[\alpha_{k\lambda}(t) e^{i\vec{k}\cdot\vec{r}} \hat{\varepsilon}_{k\lambda} + \alpha_{k\lambda}^{+}(t) e^{-i\vec{k}\cdot\vec{r}} \hat{\varepsilon}_{k\lambda}^{*} \right] \tag{9}$$

where the discrete summation runs over two polarizations λ, circular left and right, and $\hat{\varepsilon}_{k\lambda}$ is the polarization complex unit vector.

According to the density of states theory, the continuous summation over the modes k in (9) can be transformed to a discrete one over the k-mode photons by considering the electromagnetic field in a cavity of volume V, so that we can put

$$\int \frac{d^3k}{(2\pi)^{3/2}} \to \sqrt{\frac{1}{V}} \sum_{k} \tag{10}$$

Here, it is extremely important to note that the last relation is only valid for an ensemble of k-mode photons, whose wavelengths λ_k are much shorter than the characteristic dimensions of the volume V:

$$\lambda_k \ll V^{1/3} \quad (\forall\ k) \tag{11}$$

Adopting now Heisenberg's representation

$$\alpha_{k\lambda}(t) = \alpha_{k\lambda} e^{-i\omega_k t}; \quad \alpha_{k\lambda}^{+}(t) = \alpha_{k\lambda}^{+} e^{i\omega_k t} \tag{12}$$

and using the transformation relation (10), the vector potential of the electromagnetic field is written in QED as [23, 28]

$$\vec{A}(\vec{r},t) = \sum_{k} \sqrt{\frac{\hbar}{2\varepsilon_0 \omega_k V}} \sum_{\lambda=1}^{2} \left[\alpha_{k\lambda} e^{i(\vec{k}\cdot\vec{r}-\omega_k t)} \hat{\varepsilon}_{k\lambda} + \alpha_{k\lambda}^{+} e^{-i(\vec{k}\cdot\vec{r}-\omega_k t)} \hat{\varepsilon}_{k\lambda}^{*} \right] \tag{13}$$

where the vector potential amplitude for each mode k is defined exactly as in Eq. (8).

Considering the scalar potential of the electromagnetic field to be constant in space, then the electric field is written as

$$\vec{E}(\vec{r},t) = -\partial \vec{A}(\vec{r},t)/\partial t = i \sum_{k} \sqrt{\frac{\hbar \omega_k}{2\varepsilon_0 V}} \times \sum_{\lambda=1}^{2} \left[\alpha_{k\lambda} e^{i(\vec{k}\cdot\vec{r}-\omega_k t)} \hat{\varepsilon}_{k\lambda} - \alpha_{k\lambda}^{+} e^{-i(\vec{k}\cdot\vec{r}-\omega_k t)} \hat{\varepsilon}_{k\lambda}^{*} \right] \tag{14}$$

The amplitudes in (13) and (14) have been obtained using the density of states theory and are valid only on the condition (11).

Now, the boundary conditions in cavities and waveguides impose the wave vectors k of the modes to be higher than a characteristic cut-off value $k > k_{\text{cut-off}}$ ($\lambda_k < \lambda_{\text{cut-off}}$), depending on the dimensions as well as on the shape of the volume V containing the radiation field [42]. Consequently, for a volume V with finite dimensions, the summation in (13) and (14) runs only over the modes k with wave vectors higher than the minimum cut-off value $k_{\text{cut-off}}(V)$ corresponding to the shape and dimensions of V, so that we can write more precisely

$$\vec{A}(\vec{r},t) = \sum_{k>k_{\text{cut-off}}(V)} \sqrt{\frac{\hbar}{2\varepsilon_0 \omega_k V}} \times \sum_{\lambda=1}^{2} \left[\alpha_{k\lambda} e^{i(\vec{k}\cdot\vec{r}-\omega_k t)} \hat{\varepsilon}_{k\lambda} + \alpha_{k\lambda}^{+} e^{-i(\vec{k}\cdot\vec{r}-\omega_k t)} \hat{\varepsilon}_{k\lambda}^{*} \right] \tag{15}$$

$$\vec{E}(\vec{r},t) = i \sum_{k>k_{\text{cut-off}}(V)} \sqrt{\frac{\hbar\omega_k}{2\varepsilon_0 V}}$$

$$\times \sum_{\lambda=1}^{2} \left[\alpha_{k\lambda} e^{i(\vec{k}\cdot\vec{r}-\omega_k t)} \hat{\varepsilon}_{k\lambda} - \alpha_{k\lambda}^{+} e^{-i(\vec{k}\cdot\vec{r}-\omega_k t)} \hat{\varepsilon}_{k\lambda}^{*} \right] \quad (16)$$

The last equations represent the vector potential and the electric field of a large number of k-mode photons in a finite volume V with $\lambda_k \ll V^{1/3}$ ($\forall\, k$).

Following the last relations, we can now deduce the amplitude of the electric field $|\vec{\varepsilon}_k|$ and magnetic field $|\vec{\beta}_k|$ of a single k-mode photon:

$$|\vec{\varepsilon}_k| = \sqrt{\frac{\hbar\omega_k}{2\varepsilon_0 V}} \quad \text{and} \quad |\vec{\beta}_k| = \sqrt{\frac{\mu_0 \hbar\omega_k}{2V}} \quad (17)$$

These expressions are sometimes used in the literature for the free single photon electric and magnetic field amplitudes. However, as mentioned above, the expressions (15) and (16) are only valid for a large number of photons in a cavity V under the condition (11), and consequently, the attribution of the amplitudes (17) to free-of-cavity single photons should be considered with caution.

3. Photon wave function from the electric and magnetic fields

In quantum theory, a wave function $\Psi(\vec{r},t)$, depending on space and time, characterizes precisely the quantum state of a given particle. It satisfies Schrödinger's equation with eigenvalue the corresponding energy of the quantum state. However, in the case of the single photon, which is a relativistic massless "particle," the wave functions that have been proposed until now are not in general satisfactory. Nevertheless, some very fruitful studies have already been carried out in order to establish a photon wave function [43–48]. The most pertinent theoretical developments are based on the complex vector function $\vec{F}(\vec{r},t)$ initially introduced by Riemann:

$$\vec{F}(\vec{r},t) = \frac{1}{2\sqrt{\varepsilon_0}}\vec{D}(\vec{r},t) + i\frac{1}{2\sqrt{\mu_0}}\vec{B}(\vec{r},t) \tag{18}$$

where the electric displacement flux density $\vec{D}(\vec{r},t)$ and the magnetic field flux density $\vec{B}(\vec{r},t)$ are defined, respectively, through the electric and magnetic field intensities $\vec{E}(\vec{r},t)$ and $\vec{H}(\vec{r},t)$ following the expressions

$$\vec{D}_{\alpha\beta}(t) = \hat{\varepsilon}_{\alpha\beta}\cdot\vec{E}_{\alpha\beta}(t); \quad \vec{B}_{\alpha\beta}(t) = \hat{\mu}_{\alpha\beta}\cdot\vec{H}_{\alpha\beta}(t) \tag{19}$$

with the electric permittivity $\hat{\varepsilon}_{\alpha\beta}$ and magnetic permeability $\hat{\mu}_{\alpha\beta}$ being tensors ($\alpha = x, y, z$ and $\beta = x, y, z$), characterizing the intrinsic electric and magnetic nature of the medium, in which the electric and magnetic field intensities subsist.

At that point, it is important to mention that the electric field $\vec{E}(\vec{r},t)$ and the magnetic induction $\vec{B}(\vec{r},t)$ are fundamental fields, while $\vec{D}(\vec{r},t)$ and $\vec{H}(\vec{r},t)$ are fields that include macroscopically the response of the medium. In the case of an isotropic medium, in which the electric and magnetic properties are the same in all directions, the electric field is parallel to the electric displacement, while the magnetic field is parallel to the magnetic field flux intensity

$$\vec{D}(\vec{r},t) = \varepsilon\,\vec{E}(\vec{r},t) \quad \vec{H}(\vec{r},t) = \frac{1}{\mu}\,\vec{B}(\vec{r},t) \tag{20}$$

Using the complex function $\vec{F}(\vec{r},t)$, Maxwell's equations in a homogeneous medium become

$$i\frac{\partial}{\partial t}\vec{F}(\vec{r},t) = c\vec{\nabla}\times\vec{F}(\vec{r},t) \tag{21}$$

$$\vec{\nabla}\cdot\vec{F}(\vec{r},t) = 0 \tag{22}$$

In this way, a suggested photon wave function was defined as

$$\Psi(\vec{r},t) = \begin{pmatrix} \frac{1}{2\sqrt{\varepsilon_0}}\vec{D}(\vec{r},t) + i\frac{1}{2\sqrt{\mu_0}}\vec{B}(\vec{r},t) \\ \frac{1}{2\sqrt{\varepsilon_0}}\vec{D}(\vec{r},t) - i\frac{1}{2\sqrt{\mu_0}}\vec{B}(\vec{r},t) \end{pmatrix} \tag{23}$$

which is a six-component vector satisfying Schrödinger's equation in the form

$$i\hbar\frac{\partial}{\partial t}\Psi(\vec{r},t) = c\begin{pmatrix} -i\hbar\vec{\nabla}\cdot\vec{S} & 0 \\ 0 & i\hbar\vec{\nabla}\cdot\vec{S} \end{pmatrix}\Psi(\vec{r},t) \tag{24}$$

where $\vec{S}$ is the spin matrix for a spin-1 particle with the Cartesian components

$$S_x = \begin{pmatrix} 0 & 0 & 0 \\ 0 & 0 & -i \\ 0 & i & 0 \end{pmatrix} \quad S_y = \begin{pmatrix} 0 & 0 & i \\ 0 & 0 & 0 \\ -i & 0 & 0 \end{pmatrix} \quad S_z = \begin{pmatrix} 0 & -i & 0 \\ i & 0 & 0 \\ 0 & 0 & 0 \end{pmatrix} \tag{25}$$

The square modulus of the above-defined photon wave function $\Psi(\vec{r},t)$ provides the classical electromagnetic field energy density (1) at an instant t and at a given coordinate $\vec{r}$ on the propagation axis:

$$|\Psi(\vec{r},t)|^2 = W(\vec{r},t) = \frac{1}{2}\left(\varepsilon_0|\vec{E}(\vec{r},t)|^2 + \frac{1}{\mu_0}|\vec{B}(\vec{r},t)|^2\right) \tag{26}$$

Obviously, the energy can be expressed directly by integrating in space the bilinear form of the complex vector function $\vec{F}(\vec{r},t)$, obtaining

$$E = \int \vec{F}^*(\vec{r},t)\cdot\vec{F}(\vec{r},t)\,d^3r \tag{27}$$

However, the relations (26) and (27) give, respectively, the energy density and the energy of the electromagnetic field but not precisely those of a single photon state. Furthermore, in quantum mechanics, the integration in space of the square modulus of the normalized photon wave function, given by (26), should correspond to the probability of localizing a photon and not to the energy of the classical electromagnetic field representation. In a characteristic single photon quantization volume on the propagation axis, this quantity should be equal to unity and this is not really the case here for $\Psi(\vec{r},t)$.

Finally, some different versions from the one described above have been advanced for the photon wave function but without introducing real progress in the whole problematics.

4. Spatial properties of a single photon and vector potential quantization

4.1. *Lateral expansion of a single photon, theoretical and experimental facts*

The moment of inertia I_k that may be attributed to a single k-mode photon with angular frequency ω_k and wavelength λ_k is written as

$$I_k = \frac{\hbar}{\omega_k} \tag{28}$$

From the mass-energy equivalence the mass m_k that may be attributed to a k-mode photon is

$$m_k = \frac{\hbar\omega_k}{c^2} \tag{29}$$

In order to estimate the radial expansion of the photon we may assume here a homogeneous distribution of the field in the single photon state quantization volume and consider the mechanical analogue in two simple cases, that of a cylinder whose axis is along the propagation axis and that of a sphere.

For a cylinder we have

$$I_{\text{cylinder}} = \frac{1}{2} m_k r_{\text{cylinder}}^2 = \frac{1}{2}\frac{\hbar\omega_k}{c^2} r_{\text{cylinder}}^2 = \frac{\hbar}{\omega_k} \tag{30}$$

and consequently

$$r_{\text{cylinder}} = \left(\frac{2c^2}{\omega_k^2}\right)^{1/2} = \frac{\lambda_k}{\pi\sqrt{2}} \sim \frac{\lambda_k}{4.4} \tag{31}$$

In the case of a spherical distribution of the field inside the photon quantization volume, we have

$$I_{\text{sphere}} = \frac{2}{5} m_k r_{\text{sphere}}^2 = \frac{2}{5}\frac{\hbar\omega_k}{c^2} r_{\text{sphere}}^2 = \frac{\hbar}{\omega_k} \tag{32}$$

so that the sphere radius is

$$r_{\text{sphere}} = \left(\frac{5}{2}\right)^{1/2} \frac{c}{\omega_k} = \frac{\lambda_k}{\pi\sqrt{8/5}} \sim \frac{\lambda_k}{3.9} \tag{33}$$

Consequently, from the above simple theoretical calculations, we may presume that the radius of a single k-mode photon should be a fraction of its wavelength, roughly $\sim\lambda_k/4$.

From the experimental point of view, the fact that a single-mode plane electromagnetic wave should have a minimum radius was already suspected since the very first applications of the Faraday grid shielding. As mentioned in the introduction, various experiments carried out by Robinson, Hadlock, Hunter and Wadlinger [16–19] using microwaves and measuring the transmitted power through rectangular or circular apertures concluded that the photon lateral expansion should indeed be close to a quarter of its wavelength $\sim\lambda_k/4$, confirming the above simple calculations based on the moment of inertia.

It is well established [20–22] that a single photon cannot be conceived in a length shorter than its wavelength, and it cannot be localized within a volume less than $\sim\lambda_k^3$ [21, 23], entailing that the single photon quantization volume should be proportional to λ_k^3, i.e., to ω_k^{-3}.

4.2. *Single photon quantization volume*

4.2.1. *The single photon quantization volume from the density of states theory*

According to the density of states theory, taking into account the two spin values, $\pm\hbar$, corresponding to circular left and right polarization, as well as the two possible directions, $\pm z$, along the same propagation axis z, the number of k-mode photons $dn(\omega_k)$ in the quantization volume V in the frequency interval between ω_k and $\omega_k+d\omega_k$ is written as [23, 27]

$$dn(\omega_k) = 4V\frac{\omega_k^2}{\pi^2 c^3}d\omega_k \tag{34}$$

where we have also considered that all the possible states in the volume V are occupied.

Thus,

$$n(\omega_k) = 4V\frac{\omega_k^3}{3\pi^2 c^3} \tag{35}$$

Considering now that $n(\omega_k) = 1$, the corresponding quantization volume that could be attributed to a single photon is

$$V_k = \left(\frac{3}{4}\pi^2 c^3\right) \omega_k^{-3} \tag{36}$$

The last expression is in agreement with the experimental evidence that the single photon quantization volume is proportional to ω_k^{-3}, obtaining

$$V_k = \frac{3}{32\pi}\lambda_k^3 \simeq 0.029\,\lambda_k^3 \tag{37}$$

Thus, the minimum quantization volume corresponding to a single photon with a wavelength λ_k is roughly 3% of the volume λ_k^3.

4.2.2. *The single photon quantization volume from the energy normalization*

We can also obtain the single photon quantization volume in a different way by normalizing the electromagnetic energy of a plane-wave mode to that of a single photon:

$$E_k = \int \varepsilon_0 \omega_k^2 \alpha_{0k}^2(\omega_k) \left[\hat{\varepsilon}_{k\lambda} e^{i(\vec{k}\cdot\vec{r}-\omega_k t+\phi)} + \hat{\varepsilon}_{k\lambda}^* e^{-i(\vec{k}\cdot\vec{r}-\omega_k t+\phi)}\right]^2 d^3r = \hbar\omega_k \tag{38}$$

where $\alpha_{0k}(\omega_k)$ represents the single k-mode photon vector potential amplitude and $\hat{\varepsilon}_{k\lambda}$ the polarization complex unit vector.

The last equation holds at any instant t if the k-mode photon polarization unit vector $\hat{\varepsilon}_{k\lambda}$ has two orthogonal components, $\hat{e}_1$ and $\hat{e}_2$, such as $\hat{e}_1 \cdot \hat{e}_2 = 0$ and $\hat{\varepsilon}_{k\lambda} = \sigma_{k1}\hat{e}_1 + \sigma_{k2}\hat{e}_2$ with $|\sigma_{k1}|^2 + |\sigma_{k2}|^2 = 1$. Obviously, these conditions are satisfied naturally by circular left (L) and right (R) polarization unit vectors $\hat{\varepsilon}_{L,R} = \frac{1}{\sqrt{2}}(\hat{e}_1 \pm i\hat{e}_2)$, so that we have

$$E_k = \int 2\varepsilon_0 \omega_k^2\, \alpha_{0k}^2(\omega_k) d^3r = \hbar\omega_k \tag{39}$$

Now, the dimension analysis of the vector potential issued from the general solution of Maxwell's equations yields that it is proportional to an angular frequency [3–6]

$$\vec{A}(\vec{r},t) = \frac{\mu}{4\pi}\int \frac{\vec{J}(\vec{r'}, t - |\vec{r} - \vec{r'}|/c)}{|\vec{r} - \vec{r'}|} d^3 r \propto \omega \tag{40}$$

where μ is the magnetic permeability and J is the current density ($Cm^{-2}s^{-1}$).

Thus, considering in (5) that the vector potential amplitude is proportional to ω, then $\langle W_k \rangle$ is proportional to ω^4, which is confirmed experimentally for the energy density radiated by a dipole [5,6]. The concluded property for the vector potential amplitude is related to its natural units and consequently is gauge-independent.

Hence, we may consider that the vector potential amplitude $\alpha_{0k}(\omega_k)$ of a single k-mode photon can be expressed as follows [49–51]:

$$\alpha_{0k}(\omega_k) = \xi \omega_k \tag{41}$$

where ξ is a constant to be determined through the normalization procedure.

Obviously, when introducing (41) in (39), an appropriate volume V_k has to be considered for the single k-mode photon for the equation to hold:

$$E_k = 2\varepsilon_0 \xi^2 \omega_k^4 V_k = \hbar \omega_k \tag{42}$$

From the last relation, we deduce that the energy density of a single photon state is

$$W_k = 2\varepsilon_0 \xi^2 \omega_k^4 \tag{43}$$

We also deduce from (42) that the characteristic volume of a cavity-free single photon is written as

$$V_k = \left(\frac{\hbar}{2\varepsilon_0 \xi^2}\right) \omega_k^{-3} \tag{44}$$

which is in agreement with the experimental evidence regarding the frequency dependence.

Considering the propagation along the z-axis, with ρ the lateral coordinate, and respecting the experimental results by imposing the limits $0 \leq z \leq \lambda_k$ and $0 \leq \rho \leq \eta\lambda_k$ [49], where η characterizes the radial extension, Eq. (39) gives

$$\varepsilon_0 \xi^2 \eta^2 \omega_k^4 \lambda_k^3 = (2\pi c)^3 \varepsilon_0 \xi^2 \eta^2 \omega_k = \hbar\omega_k \tag{45}$$

Hence,

$$\eta\xi = \pm \frac{1}{(2\pi)^{3/2}} \sqrt{\frac{\hbar}{\varepsilon_0 c^3}} \tag{46}$$

Instead of using the approximate experimental value of $\eta \sim 1/4$, we can now try to get a more precise result. In fact, Eq. (45) is written with a slight rearrangement as

$$4\pi\, c \left(\frac{1}{4\pi\, c} \varepsilon_0 \omega_k^2 \eta^2 \alpha_{0k}(\omega_k) \lambda_k^3 \right) \alpha_{0k}(\omega_k) = 4\pi\, c\, Q\, \alpha_{0k}(\omega_k) = \hbar\omega_k \tag{47}$$

where Q has charge units. Using (41) and (46), we get

$$Q = \frac{1}{4\pi\, c} \varepsilon_0 \omega_k^2 \eta^2 \alpha_0(\omega_k) \lambda_k^3 = \pm \frac{\eta}{2} \sqrt{\varepsilon_0 h\, c} \tag{48}$$

with h being Planck's constant.

It is of high importance to underline here that when introducing in (48) the approximate experimental value for $\eta \sim 1/4$, we get

$$Q = \pm 1.6\ 10^{-19} C \tag{49}$$

which is the electron–positron charge, a physical constant appearing naturally [51] in the normalization process.

We deduce that the physical origin of the elementary charge is strongly related to the photon vector potential [51, 53]. Now, we can define more precisely η and ξ by replacing Q in (48) by the elementary charge e, and using the fine structure constant $\alpha = e^2/4\pi\varepsilon_0\hbar c = 1/137$, we get

$$\eta = \sqrt{8\alpha} \tag{50}$$

and from Eq. (46), we obtain the vector potential amplitude normalization constant [50, 51, 53]

$$\xi = \pm\frac{1}{(2\pi)^{3/2}}\sqrt{\frac{\hbar}{8\alpha\varepsilon_0 c^3}} = \pm\frac{\hbar}{4\pi|e|c} = \pm 1.747\ 10^{-25}\,\mathrm{Volt\,m^{-1}s^2} \tag{51}$$

The single k-mode photon quantization volume is now written as

$$V_k = \left(\frac{\hbar}{2\varepsilon_0\xi^2}\right)\omega_k^{-3} = \frac{\eta^2}{2}\lambda_k^3 = 4\alpha\lambda_k^3 \simeq 0.029\,\lambda_k^3 \tag{52}$$

which is equivalent to (37).

Consequently, the helicoidally distributed vector potential field of a single photon state over a wavelength occupies roughly 3% of the λ_k^3 volume, as we have seen before, and has a wave front section [54]:

$$S_k = 8\pi\alpha\lambda_k^2 \simeq 0.18\,\lambda_k^2 \tag{53}$$

The last relations are useful for technological applications, such as wave guides and fiber optic transmissions or Faraday grid shielding.

Thus, the fundamental properties of the photon, energy E_k, momentum $\vec{p}_k$ and wave vector $\vec{k}$, are complemented by the vector potential amplitude α_{0k} expressing its intrinsic electromagnetic nature

$$E_k/\hbar = |\vec{p}_k|c/\hbar = |\vec{k}|c = \alpha_{0k}/|\xi| = \omega_k \tag{54}$$

From the relation (54) and Heisenberg's *energy–time* uncertainty, a *vector potential–time* uncertainty is drawn:

$$\delta E_k\,\delta t \geq \hbar \rightarrow \delta\alpha_{0k}\delta t \geq |\xi| \tag{55}$$

showing that the particle (energy) and the wave (vector potential) are intrinsic properties of the single photon nature.

5. Single photon wave–particle (classical–quantum) physical properties

Following the above considerations, the vector potential for a free single k-mode photon with λ-polarization (circular left or right) is written in both classical and quantum formalisms, respectively, as [53]

$$\begin{aligned}\tilde{\alpha}_{k\lambda}(\vec{r},t) &= \xi\omega_k\left[\hat{\varepsilon}_{k\lambda}\alpha_{k\lambda}e^{i\left(\vec{k}\cdot\vec{r}-\omega_k t\right)} + \hat{\varepsilon}^*_{k\lambda}\alpha^+_{k\lambda}e^{-i\left(\vec{k}\cdot\vec{r}-\omega_k t\right)}\right]\\ \vec{\alpha}_{k\lambda}(\vec{r},t) &= \xi\omega_k\left[\hat{\varepsilon}_{k\lambda}e^{i\left(\vec{k}\cdot\vec{r}-\omega_k t\right)} + \hat{\varepsilon}^*_{k\lambda}e^{-i\left(\vec{k}\cdot\vec{r}-\omega_k t\right)}\right]\end{aligned} \tag{56}$$

For a free k-mode photon, the volume V_k expressed by (37) and (52), corresponds to the space in which the quantized vector potential rotates around the propagation axis at the angular frequency ω_k over a period (i.e., over a wavelength λ_k). This precession generates orthogonal electric and magnetic fields, whose amplitudes are

$$|\vec{\varepsilon}_k| = |-\partial\vec{\alpha}_{k\lambda}(\vec{r},t)/\partial t| \propto |\xi|\omega_k^2; \quad |\vec{\beta}_k| \propto \sqrt{\varepsilon_0\mu_0}\,|\xi|\,\omega_k^2 \tag{57}$$

It is important to note that the last expressions are directly proportional to the square of the angular frequency and do not depend on any external volume parameter [53, 55].

The quantum properties of a single photon, energy, momentum and spin result readily from the integration of the classical electromagnetic expressions over the quantization volume V_k, linking in this way the classical (wave) to the quantum (particle) representations [51, 53, 54]. Indeed, using (41) and (52), the energy is written as

$$E_k = \int_{V_k} 2\varepsilon_0\alpha_{0k}^2\omega_k^2 d^3r = \int_{V_k} 2\varepsilon_0\xi^2\omega_k^4 d^3r = 2\varepsilon_0\xi^2\omega_k^4 V_k = \hbar\omega_k \tag{58}$$

Considering circular polarizations for the electric and magnetic field components, and using (57), the momentum is obtained directly:

$$\vec{p}_k = \int_{V_k} \varepsilon_0\,\vec{\varepsilon}_{k\lambda}\times\vec{\beta}_{k\lambda}d^3r = \varepsilon_0(\sqrt{2}\omega_k\alpha_{0k})(\sqrt{2}\omega_k\alpha_{0k}/c)\,V_k\vec{k}/|\vec{k}| = \hbar\vec{k} \tag{59}$$

According to the classical electromagnetic theory, the spin is written through the electric and magnetic fields components, hence, using again the circular polarization, we get

$$|\vec{S}| = \int_{V_k} \varepsilon_0 |\vec{r} \times (\vec{\varepsilon}_{k\lambda} \times \vec{\beta}_{k\lambda})| d^3 r = \varepsilon_0 (c/\omega_k)(\sqrt{2}\omega_k \alpha_{0k}) \times (\sqrt{2}\omega_k \alpha_{0k}/c)\, V_k = \hbar \tag{60}$$

where we have considered the mean value of the distance [56] for a single photon state $< |\vec{r}| >_k = c/\omega_k$.

The last relations show that the quantum properties of the single photon can be obtained from the classical electromagnetic field considered in the quantization volume V_k, entailing that the photon has naturally a three-dimensional extension, and consequently, the term "wave–corpuscle" should be more appropriate.

Thus, a photon is not a point particle, and Heisenberg's uncertainty relation for the position and momentum is precisely due to its spatial properties and more precisely to the quantization volume V_k. Indeed, replacing V in (8) by V_k, we get the single photon vector potential amplitude operators expressed through the annihilation and creation operators [51, 53]

$$\begin{aligned} \tilde{\alpha}_{0k\lambda} &= \xi \omega_k \alpha_{k\lambda} \\ \tilde{\alpha}^*_{0k\lambda} &= \xi \omega_k \alpha^+_{k\lambda} \end{aligned} \tag{61}$$

The corresponding position $\tilde{Q}_{k\lambda}$ and momentum $\tilde{P}_{k\lambda}$ Hermitian operators [23,27,28] are written as

$$\tilde{Q}_{k\lambda} = \sqrt{\varepsilon_0 V_k}(\tilde{\alpha}_{0k\lambda} + \tilde{\alpha}^*_{0k\lambda}); \quad \tilde{P}_{k\lambda} = -i\omega_k \sqrt{\varepsilon_0 V_k}(\tilde{\alpha}_{0k\lambda} - \tilde{\alpha}^*_{0k\lambda}) \tag{62}$$

Thus, introducing (61) in (62) and using (52), Heisenberg's commutation relation, a fundamental concept in quantum theory, results directly in

$$\begin{aligned} \left[\tilde{Q}_{k\lambda}, \tilde{P}_{k'\lambda'}\right] &= -i\varepsilon_0 \omega^2_{k'} \omega_k \sqrt{V_k V_{k'}} \left[(\xi a_{k\lambda} + \xi\, a^+_{k\lambda}),\ (\xi a_{k'\lambda'} - \xi a^+_{k'\lambda'})\right] \\ &= i\hbar \delta_{kk'} \delta_{\lambda\lambda'} \end{aligned} \tag{63}$$

The position–momentum uncertainty as well as the energy and vector potential uncertainties with respect to time are intrinsic physical properties of the nature of the photon.

In this way, the general equation for the vector potential of the electromagnetic wave considered as a superposition of plane wave modes is written as

$$\vec{A}(\vec{r},t) = \sum_{k,\lambda} \xi\omega_k \left[\hat{\varepsilon}_{k\lambda} e^{i\left(\vec{k}\cdot\vec{r}-\omega_k t+\phi\right)} + \hat{\varepsilon}^*_{k\lambda} e^{-i\left(\vec{k}\cdot\vec{r}-\omega_k t+\phi\right)}\right] \tag{64}$$

and that of a large number of cavity-free photons in QED is

$$\tilde{A} = \sum_{k,\lambda} \xi\omega_k \left[a_{k\lambda}\hat{\varepsilon}_{k\lambda} e^{i(\vec{k}\cdot\vec{r}-\omega_k t+\phi)} + a^+_{k\lambda}\hat{\varepsilon}^*_{k\lambda} e^{-i(\vec{k}\cdot\vec{r}-\omega_k t+\phi)}\right] \tag{65}$$

In what follows, the photon wave–particle equation is introduced, showing that the vector potential with the quantized amplitude can naturally be a wave function for the photon.

6. Wave–particle equation for the photon and vector potential wave function

It can be readily demonstrated that the vector potential $\vec{\alpha}_{k\lambda}(\vec{r},t)$ with the quantized amplitude $\xi\omega_k$ expressed by (41) satisfies Maxwell's wave propagation

$$\vec{\nabla}^2_k \vec{\alpha}_{k\lambda}(\vec{r},t) - \frac{1}{c^2}\frac{\partial^2}{\partial t^2}\vec{\alpha}_{k\lambda}(\vec{r},t) = 0 \tag{66}$$

and the photon *vector potential–energy* (*wave–particle*) equation

$$i\begin{pmatrix}\xi\\ \hbar\end{pmatrix}\frac{\partial}{\partial t}\vec{\alpha}_{k\lambda}(\vec{r},t) = \begin{pmatrix}\tilde{\alpha}_{0k}\\ \tilde{H}_k\end{pmatrix}\vec{\alpha}_{k\lambda}(\vec{r},t) \tag{67}$$

where the operator $\vec{\nabla}_k$ acts upon the mode k.

Obviously, the relativistic Hamiltonian for a massless particle $\tilde{H}_k = -i\hbar c\vec{\nabla}_k$ as well as the vector potential amplitude operator $\tilde{\alpha}_{0k} = -i\xi c\vec{\nabla}_k$, respectively, have the eigenvalues $\hbar\omega_k$ and $\xi\omega_k$ [51, 53–55, 57–59]. It is worth noting the symmetry between $\{E_k, \hbar\}$ and $\{\alpha_{0k}, \xi\}$ characterizing, respectively, the particle (energy) and electromagnetic wave (vector potential) intrinsic natures of a single photon.

Consider now the emission in vacuum of a single k-mode photon at the coordinate $\vec{r}$ at an instant t. The probability of being localized at the instant t' at the coordinate $\vec{r}' = \vec{r} + c(t' - t)\hat{r}$ can be obtained by the square of the modulus of the vector potential function $\vec{\alpha}_{k\lambda}(\vec{r}, t)$:

$$|\vec{\alpha}_{k\lambda}(z,t)|^2 = \xi^2 \omega_k^2 \propto \lambda_k^{-2} \tag{68}$$

It is evident that the higher the frequency, the shorter the photon wavelength and the higher the localization probability. This is in agreement with Heisenberg's uncertainty principle as well as with the experimental evidence, and consequently, the vector potential function $\vec{\alpha}_{k\lambda}(\vec{r}, t)$ can be considered as a real wave function for the photon that can be suitably normalized.

The photon momentum according to (59) is $\vec{p}_k = \hbar\vec{k} = (h/\lambda_k)\,\hat{k}$ and considering the propagation along the z-axis, the Heisenberg position–momentum uncertainty, expressed by (63), is written now simply as

$$\delta z\ \delta p_{k_z} \geq h\ \rightarrow \delta z\ \delta(1/\lambda_k) \geq 1 \tag{69}$$

The physical meaning of the last relation is that the localization uncertainty of a single photon on the propagation axis is of the order of the wavelength λ_k. Furthermore, as already mentioned, a single photon is localizable only within a volume proportional to λ_k^3.

Now, taking into account the above considerations and weighting the vector potential function $\vec{\alpha}_{k\lambda}(\vec{r}, t)$ by the factor $(\varepsilon_0\omega_k/\hbar)^{1/2}$, a wave function for the free photon can be defined as

$$\Phi_{k,(L,R)}(\vec{r}, t) = \left(\frac{\varepsilon_0\omega_k}{\hbar}\right)^{1/2} \left[\alpha_{0k}(\omega_k)\left(\hat{\varepsilon}_{k,(L,R)} e^{i(\vec{k}\cdot\vec{r} - \omega_k t)} + \hat{\varepsilon}^*_{k,(L,R)} e^{-i(\vec{k}\cdot\vec{r} - \omega_k t)}\right)\right] \tag{70}$$

where the polarization can be either left (L) or right (R) circular.

Whatever the circular polarization, the square of the modulus of the normalized general wave function (70) gives the inverse of the single photon quantization volume (52):

$$\begin{aligned} \left|\Phi_{k,(L,R)}(\vec{r}, t)\right|^2 &= \frac{\varepsilon_0\omega_k}{\hbar}\left[\xi^2\omega_k^2\left(\left|\hat{\varepsilon}_{k,(L,R)}\right|^2 + \left|\hat{\varepsilon}^*_{k,(L,R)}\right|^2\right)\right] \\ &= \frac{2\varepsilon_0\xi^2\omega_k^3}{\hbar} = \frac{1}{V_k} \end{aligned} \tag{71}$$

so that

$$\int_{V_k} \left|\Phi_{k,(L,R)}(\vec{r},t)\right|^2 d^3r = 1 \tag{72}$$

The probability of finding the photon in the volume V_k around the coordinate $\vec{r}$ on the propagation axis equals one.

Consequently, the photon wave function defined in (70), obtained from the vector potential quantization at a single photon level, satisfies the propagation equation (66), the vector potential–energy equation (67), the normalized probability condition (72) and represents a real eigenstate of the relativistic massless particle Hamiltonian and the corresponding momentum operator with eigenvalues $\hbar\omega_k$ and $\hbar|\vec{k}|$, respectively.

7. Conclusion

Based on the experimental confirmation that the vector potential is a real physical entity, we advanced theoretical developments complementing the standard formalism for the description of a single photon and the definition of a wave function.

The quantization of the vector potential amplitude of a cavity-free k-mode photon with angular frequency ω_k is $\alpha_{0k} = \xi\omega_k$, where $\xi = \hbar/4\pi ec$, and yields naturally a *vector potential-energy* (electromagnetic wave–particle) equation, expressing the simultaneous wave–particle nature of the photon.

A single photon is an indivisible segment of the electromagnetic field, extending over a wavelength λ_k and composed of the quantized vector potential rotating at the angular frequency ω_k with circular left or right polarization corresponding, respectively, to spin $\pm\hbar$. The precession of the vector potential gives birth to orthogonally oscillating electric and magnetic fields, whose amplitudes are proportional to the square of the angular frequency $|\xi|\,\omega_k^2$. The lateral expansion of the photon has been estimated experimentally, yielding a photon quantization volume V_k proportional to λ_k^3. The energy and momentum of the photon as well as the spin are obtained from the classical electromagnetic expressions integrated over the volume V_k, showing that the photon is not a point particle but a “wave–corpuscle.”

Indeed, it is readily demonstrated that the origin of Heisenberg's uncertainty lies precisely in the photon spatial properties.

The quantized vector potential $\Phi_{k,(L,R)}(\vec{r},t)$, with circular left (L) or right (R) polarization, defined in Eq. (70), satisfies both the propagation equation and the *vector potential-energy* equation, behaving as a natural wave function for the photon. In fact, the square modulus of its amplitude is proportional to the square of the angular frequency and consequently can be employed for defining a localization probability, yielding that the higher the frequency, the better the localization, in agreement with the experiments. Furthermore, the normalization factor $(\varepsilon_0\omega_k/\hbar)^{1/2}$ ensures that the summation of $|\Phi_{k,(L,R)}(\vec{r},t)|^2$ over the quantization volume V_k equals unity.

Thus, a photon appears to be a three-dimensional electromagnetic "wave–corpuscle," carrying the energy quantum $\hbar\omega_k$ and the momentum $\hbar|\vec{k}|$, guided by the non-local vector potential function $\Phi_{k,(L,R)}(\vec{r},t)$ with circular polarization corresponding to spin $\pm\hbar$.

References

[1] I. Newton, *Opticks* (Dover, New York, NY, USA, 1952).

[2] V. Ronchi, *The Nature of Light: An Historical Survey* (Harvard University Press, Cambridge, MA, USA, 1970).

[3] M. Born and E. Wolf, *Principles of Optics* (Cambridge University Press, Cambridge, UK, 1999).

[4] B. E. A. Saleh and M. C. Teich, *Fundamentals of Photonics* (John Wiley & Sons, Hoboken, NJ, USA, 2007).

[5] F. H. Read, *Electromagnetic Radiation* (John Wiley & Sons, New York, NY, USA, 1980).

[6] J. D. Jackson, *Classical Electrodynamics*, 3rd ed. (Wiley, New York, NY, USA, 1998).

[7] M. Planck, *The Theory of Heat Radiation* (Dover, New York, NY, USA, 1959).

[8] B. H. Bransden and C. J. Joachain, *Physics of Atoms and Molecules*, 2nd ed. (Pearson, London, UK, 2003).

[9] H. Hertz, Ueber einen Einfluss des ultravioletten Lichtes auf die electrische Entladung, *Ann. Phys.* 267, 983 (1887).

[10] A. Einstein, *The Collected Papers of Albert Einstein*, J. Stachel, D. C. Cassidy, J. Renn, and R. Schulmann, Eds. (Princeton University Press, Princeton, NJ, USA, 1987).

[11] A. H. Compton, A quantum theory of the scattering of X-rays by light elements, *Phys. Rev.* 21, 483 (1923).
[12] G. Wentzel, Zur Theorie des photoelektrischen Effekts, *Z. Phys.* 40, 574 (1926).
[13] G. Beck, Zur Theorie des Photoeffekts, *Z. Phys.* 41, 443 (1927).
[14] W. E. Lamb, Jr. and M. O. Scully, in *Polarisation, Matière et Rayonnement, Volume Jubilaire en l'Honneur d'Alfred Kastler* (French Physical Society, Presses Universitaires de France, Paris, France, 1969).
[15] O. Klein and Y. Nishina, Über die Streuung von Strahlung durch freie Elektronen nach der neuen relativistischen Quantendynamik von Dirac, *Z. Phys.* 52, 853 (1929).
[16] H. L. Robinson, Diffraction patterns in circular apertures less than one wavelength in diameter, *J. Appl. Phys.* 24, 35 (1953).
[17] R. K. Hadlock, Diffraction patterns at the plane of a slit in a reflecting screen, *J. Appl. Phys.* 29, 918 (1958).
[18] G. Hunter and R. L. P. Wadlinger, Photons and neutrinos as electromagnetic solitons, *Phys. Essays* 2, 158 (1989).
[19] G. Hunter, in *Quantum Uncertainties: Recent and Future Experiments and Interpretations*, W. M. Honig, D. W. Kraft, and E. Panarella, Eds., NATO ASI Series (Series B: Physics), Vol. 162 (Springer, Boston, MA, USA, 1987), p. 331.
[20] L. Mandel, Configuration-space photon number operators in quantum optics, *Phys. Rev.* 144, 1071 (1966).
[21] R. L. Pfleegor and L. Mandel, Interference of independent photon beams, *Phys. Rev.* 159, 1084 (1967).
[22] A. I. Akhiezer and V. B. Berestetskii, *Quantum Electrodynamics* (Interscience, New York, NY, USA, 1965).
[23] G. Grynberg, A. Aspect, and C. Fabre, *Introduction to Quantum Optics* (Cambridge University Press, Cambridge, UK, 2010).
[24] G. Auletta, *Foundations and Interpretation of Quantum Mechanics* (World Scientific, Singapore, 2001).
[25] A. A. Michelson *et al.*, Conference on the Michelson-Morley experiment, *Astrophys. J.* 68, 341 (1928).
[26] M. H. Mittleman, *Introduction to the Theory of Laser-Atom Interactions* (Plenum Press, New York, NY, USA, 1982).
[27] M. Weissbluth, *Photon-Atom Interactions* (Academic Press, Boston, MA, USA, 1988).
[28] J. C. Garrison and R. Y. Chiao, *Quantum Optics* (Oxford University Press, Oxford, UK, 2008).
[29] D. N. Klyshko, A. N. Penin, and B. F. Polkovnikov, Parametric luminescence and light scattering by polaritons, *JETP Lett.* 11, 5 (1970).

[30] D. C. Burnham and D. L. Weinberg, Observation of simultaneity in parametric production of optical photon, *Phys. Rev. Lett.* 25, 84 (1970).
[31] P. Ghose and D. Home, The two-prism experiment and wave-particle duality of light, *Found. Phys.* 26, 943 (1996).
[32] P. Grangier, G. Roger, and A. Aspect, Experimental evidence for a photon anticorrelation effect on a beam splitter: A new light on single-photon interferences, *Europhys. Lett.* 1, 173 (1986).
[33] C. A. Kocher and E. D. Commins, Polarization correlation of photons emitted in an atomic cascade, *Phys. Rev. Lett.* 18, 575 (1967).
[34] S. J. Freedman and J. F. Clauser, Experimental test of local hidden-variable theories, *Phys. Rev. Lett.* 28, 938 (1972).
[35] A. Aspect *et al.*, Experimental tests of realistic local theories via Bell's theorem, *Phys. Rev. Lett.* 47, 460 (1981).
[36] W. Ehrenberg and R. E. Siday, The refractive index in electron optics and the principles of dynamics, *Proc. Phys. Soc. London, Sect. B* 62, 8 (1949).
[37] R. G. Chambers, Shift of an electron interference pattern by enclosed magnetic flux, *Phys. Rev. Lett.* 5, 3 (1960).
[38] A. Tonomura *et al.*, Observation of the Aharonov–Bohm effect by electron holography, *Phys. Rev. Lett.* 48, 1443 (1982).
[39] N. Osakabe *et al.*, Experimental confirmation of Aharonov-Bohm effect using a toroidal magnetic field confined by a superconductor, *Phys. Rev. A* 34, 815 (1986).
[40] L. H. Ryder, *Quantum Field Theory* (Cambridge University Press, Cambridge, UK, 1987).
[41] P. W. Milonni, *The Quantum Vacuum: An Introduction to Quantum Electrodynamics* (Academic Press, San Diego, CA, USA, 1994).
[42] D. M. Pozar, *Microwave Engineering*, 3rd ed. (John Wiley & Sons, Hoboken, NJ, USA, 2005).
[43] J. E. Sipe, Photon wave functions, *Phys. Rev. A* 52, 1875 (1995).
[44] I. Bialynicki-Birula, Photon wave function, in *Progress in Optics*, Vol. XXXVI, E. Wolf, Ed. (Elsevier, Amsterdam, The Netherlands, 1996).
[45] B. J. Smith and M. G. Raymer, Photon wave functions, wave-packet quantization of light, and coherence theory, *New J. Phys.* 9, 414 (2007).

Chapter 7

Reappraisal of QED Fundamentals*

Abstract
In this chapter, the drawbacks of the quantum electrodynamics (QED) fundamentals are revisited and analyzed. Those concern the vector potential amplitude operator, the electromagnetic field harmonic oscillator Hamiltonian with the associated zero-point energy singularity and the electron–vacuum interaction Hamiltonian. Without stating postulates or advancing any hypothesis, it is shown that the QED difficulties can be readily overcome without compromising any of the achievements by enhancing the quantization of the vector potential amplitude to a single photon level. A photon wave function is defined as a probability amplitude normalized within an intrinsic quantization volume and satisfying Schrödinger's equation for both the photon energy and vector potential. Following the single photon vector potential quantization, the electromagnetic vacuum appears naturally, complementing the normal ordering Hamiltonian without involving singularities.

1. Introduction

> "All the 50 years of conscious brooding have brought me no closer to answer the question:
> "What are light quanta?"
> Of course, today every rascal thinks he knows the answer, but he is deluding himself."
>
> — **Albert Einstein**

*This chapter is reproduced from C. Meis. 2025. "Reappraisal of QED Fundamentals", *International Journal of Quantum Foundations,* 11(2), 2025, pp. 42–56.

QED is without any doubt among the most successful theories developed in modern physics for the description of light–matter interactions at the atomic and molecular scales. However, despite the numerous achievements and the impressive precision of some theoretical predictions compared to the experiments, Einstein's statement is valid even today because we are unable to formulate a complete and satisfactory physical description of a single photon state. Furthermore, the QED theory involves some well-known fundamental aspects, leading to ambiguities and singularities [1–3]. For many decades, the last ones had not really had a direct impact on the majority of the applications until recent astrophysical observations on the cosmic expansion have shown that the quantum vacuum energy density is finite and extremely low, contradicting the singularity of the zero-point energy in QED and in quantum field theory in general [4–10]. Hence, the well-validated astrophysical investigations generated an increasing interest in the last years toward the comprehension of the real nature of the quantum vacuum, looking for new representations overcoming the well-known singularity of the zero-point energy issued from the harmonic oscillator Hamiltonian.

On the other hand, in order to define the vector potential amplitude operators of a single photon in QED, the classical and quantum energy density equivalence has been considered, involving an external volume which is not an intrinsic photon physical property [1–3, 11]. Although the vector potential natural units are proportional to a frequency, the defined single photon vector potential amplitude in QED is proportional to the inverse square root of the frequency, yielding unphysical consequences. Thus, the dependence on frequency of the photon vector potential and that of the photon electric field are conflictual, entailing a fundamental physical incoherence. Finally, the well-known difficulties encountered in establishing a photon wave function are mainly due to the photon point particle concept, yielding major constraints for establishing a position operator [12–17].

In what follows, we first discuss extensively the origin of the above QED fundamental drawbacks, and next, we show how the enhancement of the vector potential amplitude to a single photon level permits us to remedy all these theoretical inconveniences.

2. Fundamental ambiguities in QED theory

2.1. *Photon vector potential amplitude operators*

In order to define the vector potential amplitude operators for a single k-mode photon with angular frequency ω_k, the mean value over a period of the energy density $\langle W\rangle_T$ of a monochromatic plane wave is put equal to the energy density of a single photon considered in a cavity with volume V [1,2,11]:

$$\langle W\rangle_T = 2\varepsilon_0\omega_k^2|a_{0k}(\omega_k)|^2 \equiv \frac{\hbar\omega_k}{V} \to a_{0k}(\omega_k) = \sqrt{\frac{\hbar}{2\varepsilon_0\omega_k V}} \tag{1}$$

where $\alpha_{0k}(\omega_k)$ is the vector potential amplitude of the k-mode electromagnetic plane wave, $\hbar$ is Planck's reduced constant and ε_0 is the vacuum electric permittivity.

Based on the classical–quantum energy density equivalence imposed above, the vector potential amplitude operators for a single photon are defined in QED [1,2,18] by associating the k-mode and λ-polarization photon annihilation and creation non-Hermitian operators, $a_{k\lambda}$ and $a_{k\lambda}^{+}$, respectively, to the established expression of the vector potential amplitude:

$$\tilde{a}_{k\lambda} = \sqrt{\frac{\hbar}{2\varepsilon_0\omega_k V}}a_{k\lambda}; \quad \tilde{a}_{k\lambda}^{*} = \sqrt{\frac{\hbar}{2\varepsilon_0\omega_k V}}a_{k\lambda}^{+} \tag{2}$$

We note that the left-hand side of the classical–quantum energy density equivalence in (1) is totally independent of any external parameter, while this is not the case for the right-hand side which depends on the external volume V. Here, it is extremely important to underline that the right-hand side of the relation (1) represents the energy density of the system "photon in a cavity" and not precisely that of a "single photon state."

Furthermore, the relation (1) raises another fundamental question. In fact, the general solution of the vector potential issued from Maxwell's equation is written with the well-known expression as [19,20]

$$\vec{A}(\vec{r},t) = \frac{\mu}{4\pi}\int \frac{\vec{J}(\vec{r}\,',t-|\vec{r}-\vec{r}\,'|/c)}{|\vec{r}-\vec{r}\,'|}d^3r' \tag{3}$$

where $\vec{J}$ is the current density, c is the speed of light in vacuum and μ is the magnetic permeability.

A careful unit analysis of (3) shows that the vector potential is physically proportional to a frequency. This is in contradiction with the vector potential amplitude established in Eq. (1), which is inversely proportional to the square root of the frequency.

In addition, using (1), the electric field amplitude of a single photon is written with the well-known expression as [1, 2, 18]

$$|\vec{\varepsilon}_k| = \sqrt{\frac{\hbar\omega_k}{2\varepsilon_0 V}} \tag{4}$$

Comparison between Eqs. (1) and (4) leads to the unphysical conclusion that the higher the photon frequency, the lower its vector potential amplitude and the higher its electric field. This result is conflictual and incompatible with the electromagnetic theory, entailing that the energy density equivalence (1), set roughly a century ago, has to be reappraised. We will see later that the solution to these ambiguities lies precisely in the volume parameter V.

2.2. *Origin of the harmonic oscillator Hamiltonian of the electromagnetic field*

The total energy of a multimode electromagnetic field considered in a volume V is written in classical theory as [1, 2]

$$E_{EM} = \sum_{k,\lambda} 2\varepsilon_0\omega_k^2 V |\vec{a}_{k\lambda}(\vec{r},t)|^2 \tag{5}$$

where $\vec{a}_{k\lambda}(\vec{r},t)$ is the vector potential of a single plane wave k-mode with λ-polarization.

Replacing in (5) the single k-mode photon vector potential amplitude operators defined in (2), we get the following two expressions for the quantum mechanics Hamiltonian of the electromagnetic field:

$$\tilde{H}_{\text{NO}} = \sum_{k,\lambda} a_{k\lambda}^{+} a_{k\lambda}\hbar\omega_k \to E_{\text{NO}} = \sum_{k,\lambda} n_{k\lambda}\hbar\omega_k \tag{6}$$

$$\tilde{H}_{\text{ANO}} = \sum_{k,\lambda} a_{k\lambda} a_{k\lambda}^{+}\hbar\omega_k = \sum_{k,\lambda} (a_{k\lambda}^{+} a_{k\lambda} + 1)\hbar\omega_k \to$$

$$E_{\text{ANO}} = \sum_{k,\lambda} (n_{k\lambda} + 1)\,\hbar\omega_k \tag{7}$$

where $\tilde{H}_{\mathrm{NO}}$ corresponds to the normal ordering Hamiltonian and $\tilde{H}_{\mathrm{ANO}}$ to the anti-normal ordering one. E_{NO} and E_{ANO} are the corresponding energy eigenvalues, while $n_{k\lambda}$ is the total number of k-mode and λ-polarization photons. Also, for $\tilde{H}_{\mathrm{ANO}}$, we have used the fundamental commutation relation $[a_{k\lambda}, a^{+}_{k'\lambda'}] = \delta_{kk'}\delta_{\lambda\lambda'}$.

The mean value of the last two Hamiltonians yields a harmonic oscillator Hamiltonian, which is generally introduced in QED [1,2,21] as the natural Hamiltonian of the electromagnetic field:

$$\tilde{H}_{EM} = \frac{1}{2}(\tilde{H}_{\mathrm{NO}} + \tilde{H}_{\mathrm{ANO}}) = \sum_{k,\lambda} \left(a^{+}_{k\lambda} a_{k\lambda} + \frac{1}{2} \right) \hbar\omega_k$$

$$\rightarrow E_{EM} = \sum_{k,\lambda} \left(n_{k\lambda} + \frac{1}{2} \right) \hbar\omega_k \tag{8}$$

Following this operation, the electromagnetic field is considered to be composed of single-mode harmonic oscillators corresponding to single photon states.

It is of crucial importance to mention here that we have absolutely no physical reason based on the experimental evidence to consider the mean value of the normal and anti-normal ordering Hamiltonians in order to obtain a harmonic oscillator Hamiltonian [2,21]. As we will see later, this arbitrary mathematical operation leads to unphysical results compared to the observations.

Now, a different way is also often used in order to establish a harmonic oscillator Hamiltonian for the electromagnetic field. It is based on the canonical variables of position $Q_{k\lambda}$ and momentum $P_{k\lambda}$. The vector potential amplitude and its complex conjugate can be expressed through the position and momentum variables as follows [1,2,18]:

$$a_{k\lambda} = \frac{(\omega_k Q_{k\lambda} + iP_{k\lambda})}{2\omega_k\sqrt{\varepsilon_0 V}}; \quad a^{*}_{k\lambda} = \frac{(\omega_k Q_{k\lambda} - iP_{k\lambda})}{2\omega_k\sqrt{\varepsilon_0 V}} \tag{9}$$

Using (9) in the classical electromagnetic field energy (5), we get the following expression:

$$E_{EM} = \frac{1}{2}\sum_{k,\lambda} (P^2_{k\lambda} + \omega^2_k Q^2_{k\lambda}) \pm i\omega_k [Q_{k\lambda}, P_{k\lambda}] \tag{10}$$

where the (+) sign is obtained with the normal ordering, $a^{*}_{k\lambda} a_{k\lambda}$, and the (−) one with the anti-normal ordering, $a_{k\lambda} a^{*}_{k\lambda}$.

At that level, as $Q_{k\lambda}$ and $P_{k\lambda}$ are simply canonical variables and not quantum mechanics operators, it is generally considered in the literature that the commutation in (10) vanishes, imposing $[Q_{k\lambda}, P_{k\lambda}] = 0$. In this way, a harmonic oscillator expression is obtained for the classical electromagnetic field energy:

$$E_{EM} = \frac{1}{2}\sum_{k,\lambda}(P_{k\lambda}^2 + \omega_k^2 Q_{k\lambda}^2) \tag{11}$$

Replacing in the last equation the canonical variables of position and momentum with the corresponding quantum mechanics Hermitian operators [1, 2, 18]

$$\tilde{P}_{k\lambda} = i\sqrt{\frac{\hbar\omega_k}{2}}(a_{k\lambda}^+ - a_{k\lambda}); \quad \tilde{Q}_{k\lambda} = \sqrt{\frac{\hbar}{2\omega_k}}(a_{k\lambda}^+ + a_{k\lambda}) \tag{12}$$

we get a harmonic oscillator Hamiltonian for the electromagnetic field:

$$\tilde{H}_{EM} = \sum_{k,\lambda}\left(a_{k\lambda}^+ a_{k\lambda} + \frac{1}{2}\right)\hbar\omega_k \rightarrow E_{EM} = \sum_{k,\lambda}\left(n_{k\lambda} + \frac{1}{2}\right)\hbar\omega_k \tag{13}$$

It is of high importance to underline here that, for a particle harmonic oscillator with mass $m \neq 0$ and momentum $\vec{p} = m\,d\vec{q}/dt$, with the canonical variables of position $Q = |\vec{q}|\sqrt{m}$ and momentum $P = |\vec{p}|/\sqrt{m}$, the transition from the classical expression of energy

$$E = \frac{1}{2}(P^2 + \omega^2 Q^2) \tag{14}$$

to the quantum mechanics Hamiltonian

$$\tilde{H} = \frac{1}{2}(\tilde{P}^2 + \omega^2\tilde{Q}^2) = \hbar\omega\left(a^+ a + \frac{1}{2}\right) \tag{15}$$

where $\tilde{Q} = \sqrt{\hbar/2\omega}(a^+ + a)$ and $\tilde{P} = i\sqrt{\hbar\omega/2}(a^+ - a)$, is direct and needs no commutation operations between the canonical variables P and Q [2, 22]. Consequently, for a massive particle, the harmonic oscillator Hamiltonian obtained in (15) is a direct result, expressing a real physical state, e.g., phonons in solid state physics [23].

Conversely, in the case of the electromagnetic field, a commutation operation appears between the canonical variables $Q_{k\lambda}$ and $P_{k\lambda}$ during the transition from (5) to (10). Consequently, canceling the commutation term $[Q_{k\lambda}, P_{k\lambda}]$ in (10) in order to obtain (11) just before replacing the canonical variables by the corresponding quantum mechanics operators is not mathematically correct. It is extremely important to underline here that Heisenberg's commutation relation $[\tilde{Q}_{k\lambda}, \tilde{P}_{k'\lambda'}] = i\hbar\delta_{kk'}\delta_{\lambda\lambda'}$ is a fundamental concept of quantum mechanics, which is not to be ignored when replacing classical canonical variables by the corresponding quantum mechanics operators.

Hence, keeping the commutation term $[Q_{k\lambda}, P_{k\lambda}]$ in (10) and then substituting the classical variables by the corresponding quantum operators (12), we obtain exactly the same normal ordering and anti-normal ordering Hamiltonians as in (6) and (7):

$$\begin{aligned} \tilde{H}_{EM}^{(+)} &= \frac{1}{2}\sum_{k,\lambda}(\tilde{P}_{k\lambda}^2 + \omega_k^2\tilde{Q}_{k\lambda}^2) + i\omega_k[\tilde{Q}_{k\lambda}, \hat{P}_{k'\lambda'}] \to \tilde{H}_{\mathrm{NO}} \\ &= \sum_{k,\lambda} a_{k\lambda}^{+}a_{k\lambda}\hbar\omega_k \to E_{\mathrm{NO}} = \sum_{k,\lambda} n_{k\lambda}\hbar\omega_k \end{aligned} \tag{16}$$

$$\begin{aligned} \tilde{H}_{EM}^{(-)} &= \frac{1}{2}\sum_{k,\lambda}(\tilde{P}_{k\lambda}^2 + \omega_k^2\tilde{Q}_{k\lambda}^2) - i\omega_k\left[\tilde{Q}_{k\lambda}, \tilde{P}_{k'\lambda'}\right] \to \tilde{H}_{\mathrm{ANO}} \\ &= \sum_{k,\lambda}(a_{k\lambda}^{+}a_{k\lambda} + 1)\hbar\omega_k \to E_{\mathrm{ANO}} = \sum_{k,\lambda}(n_{k\lambda} + 1)\hbar\omega_k \end{aligned} \tag{17}$$

Obviously, as quoted by many authors, the mathematical ambiguity consisting of suppressing the classical variable commutation term $[Q_{k\lambda}, P_{k\lambda}]$ just before replacing the classical variables by the non-commuting quantum mechanics operators $[\tilde{Q}_{k\lambda}, \tilde{P}_{k'\lambda'}] = i\hbar\delta_{kk'}\delta_{\lambda\lambda'}$ leads to the harmonic oscillator Hamiltonian [2,24,25]. It is revealed that considering the mean value of the normal and anti-normal ordering Hamiltonians implies inadvertently the drop of Heisenberg's uncertainty relation.

It is clear that it is impossible to get a harmonic oscillator Hamiltonian for the electromagnetic field without taking arbitrarily the mean value of the normal and anti-normal ordering Hamiltonians or without neglecting Heisenberg's uncertainty [2,21].

2.3. *The zero-point energy singularity and the physical consequences*

In the total absence of photons, i.e., for $n_{k\lambda} = 0$ in (8) or (13), we get the well-known zero-point energy of the electromagnetic field corresponding to the electromagnetic quantum vacuum:

$$E_{\text{ZPE}} = \sum_{k,\lambda} \frac{1}{2} \hbar \omega_k \tag{18}$$

whose energy density in any volume V in space is infinite:

$$\rho_{\text{ZPE}} = \frac{1}{V} \sum_{k,\lambda} \frac{1}{2} \hbar \omega_k \to \infty \tag{19}$$

This is the principal singularity in the QED theory, and the energy density of the zero-point energy is infinite at any point in space [26].

Equation (19) is a pure mathematical result, issued from the eigenstates of the harmonic oscillator Hamiltonian and should normally represent a real physical state, yielding tremendous cosmological consequences [10]. Despite this, the recent well-validated astrophysical observations upon the cosmic acceleration estimated the vacuum energy density to be finite and extremely low, roughly $\rho_{\text{ZPE}}^{\text{obs}} \sim 10^{-10}\,\text{Jm}^{-3}$, contradicting the theoretical statement resulting from the harmonic oscillator Hamiltonian [4–10]. Following the observational evidence, we are now constrained to find a finite value for the zero-point energy density.

Our knowledge of the QCD vacuum state is extremely elusive, but QED offers a readily comprehensive framework for estimating the zero-point energy density. Hence, transforming the discrete summation in (19) into a continuous one, according to the density of states theory, we obtain an integration over the frequencies, which is also infinite : [26]

$$\rho_{\text{ZPE}} = \frac{\hbar}{2\pi^2 c^3} \int_0^\infty \omega^3 d\omega \to \infty \tag{20}$$

We can now get a finite value by upper-limiting the integration. In fact, it is generally considered that the higher possible photon frequency beyond which the energy density transforms the photon into a black hole should be roughly $\sim 10^{43}$ *Hz*, corresponding to Planck's energy 10^{19} GeV. Thus, replacing the upper limit of the integration in (20) by the frequency corresponding to Planck's energy, we get a theoretical value for the electromagnetic quantum vacuum energy density $\rho_{\mathrm{ZPE}}^{\mathrm{th}} \sim 10^{110}\,\mathrm{Jm}^{-3}$. Considering the zero-point energy contribution of all the other known fields does not change radically this result, which is in disagreement with the experimental observations by 120 orders of magnitude [7, 9, 10]. This is the worst theoretical prediction in the history of science and has been called by many authors "vacuum catastrophe." Consequently, the experimental evidence implies that the zero-point energy issued from the harmonic oscillator Hamiltonian does not correspond to a real physical state.

In reality, there is no experimental evidence that a single photon is a harmonic oscillator, entailing that there is no physical reason for considering the mean value of the normal and anti-normal ordering Hamiltonians in order to obtain a harmonic oscillator Hamiltonian [2, 21]. This is also the reason why in QED the normal ordering Hamiltonian is principally used, casting aside the zero-point energy singularity.

On the other hand, it is also widespread that the zero-point energy represented by Eq. (18) is at the origin of the quantum vacuum effects, such as the spontaneous emission, the Lamb shift or even the Casimir effect. Here, we recall that the spontaneous emission effect and the Lamb shift are calculated in QED based on the properties of the single photon creation $a_{k\lambda}^{+}$ and annihilation $a_{k\lambda}$ operators without invoking at all the zero-point energy term $\sum_{k,\lambda} \frac{1}{2}\hbar\omega_k$, which is a constant and commutes with all quantum mechanics operators. Indeed, the interaction Hamiltonian between an electron and the quantum vacuum vanishes:

$$H_{\mathrm{int}} = -e\,\vec{r}\cdot\frac{i}{\hbar}\left[\tilde{A}(\vec{r},t), \sum_{k,\lambda}\frac{1}{2}\,\hbar\omega_k\right] = 0 \tag{21}$$

entailing that the zero-point energy term has absolutely no physical impact on the spontaneous emission and the Lamb shift.

Regarding the well-known Casimir effect, it has been demonstrated by many studies using different approaches [27–30] that it is physically due to Lorentz's forces or source fields and not to the zero-point energy.

Thus, as often quoted in the literature, skipping the zero-point energy by considering only the normal ordering Hamiltonian does not compromise any of the achievements related to the calculation of the vacuum effects in QED [2, 29, 30].

To conclude, the zero-point energy term $\sum_{k,\lambda} \frac{1}{2}\hbar\omega_k$ issued from the harmonic oscillator Hamiltonian of the electromagnetic field is obtained by considering, without any valuable reason, the mean value of the normal and anti-normal ordering Hamiltonians and neglecting Heisenberg's uncertainty. In addition, this term is not involved in the vacuum effects in QED and conflicts strongly with the astrophysical observational evidence. Consequently, the vacuum concept in QED has to be reappraised, and we may reasonably wonder which physical vacuum representation, depending on the creation and annihilation operators without involving singularities, could be appropriate for being associated with the normal ordering Hamiltonian.

3. Remedying QED fundamental drawbacks

In what follows, without stating postulates or axioms, we enhance the quantization of the vector potential to a single photon level, and we show that all the above ambiguities and singularities in the QED formalism are naturally overcome.

3.1. *Photon quantization volume, an intrinsic geometrical property*

It is obvious that for Eq. (1) to hold physically, the parameter V should correspond to the characteristic quantization volume of a single photon state. Such a spatial dependence of the photon may be deduced directly from the density of states theory. In fact, the number of photons $dn(\omega)$ in a given volume V in the frequency interval between ω and $\omega + d\omega$ is written as [1, 2]

$$dn(\omega) = 4V\frac{\omega^2}{\pi^2 c^3}d\omega \tag{22}$$

where we have considered the two photon spin states, corresponding classically to circular left and right helicity, and the two possible directions along the same propagation axis.

Integrating (22) implies that a quantization volume can be assigned to a single k-mode photon:

$$V_k = \left(\frac{3}{4}\pi^2 c^3\right) \omega_k^{-3} \tag{23}$$

This is compatible with the theoretical concept that a photon extends over a wavelength [2, 3, 31] and also with the experimental observations which have shown that its lateral dimension is proportional to the quarter of the wavelength [32–35], entailing that a photon can only be localized within a volume proportional to the cube of the wavelength [1–3, 18]. This experimental fact is well established since Mandel's experiments [36, 37] until now [2, 18].

Hence, introducing (23) in (1), we get directly the vector potential amplitude for a single k-mode photon:

$$\alpha_{0k}(\omega_k) = \pm\frac{1}{\pi}\sqrt{\frac{2\hbar}{3\varepsilon_0 c^3}}\omega_k = \xi\omega_k \tag{24}$$

The established photon vector potential amplitude issued from the density of states theory is now proportional to the frequency and thus compatible with the natural units of the vector potential. An equivalent result has already been deduced previously, applying a different methodology [38, 39]. It is worth noting that ξ has both positive and negative values, confirming that the photon is its own antiparticle, as always suspected in the particle theory.

Although the integration of (22) does not yield a pure integer for the photon number, the vector potential amplitude constant obtained above by the density of states theory is an excellent approximation. A precise value of ξ can be obtained by normalizing the mean energy of a single-mode plane electromagnetic wave over a period to the single photon energy [40, 41]

$$E_k = \left\langle \int \varepsilon_0 \omega_k^2 \xi^2 \omega_k^2 [\hat{\varepsilon}_{k(L.R)} e^{i(\vec{k}\cdot\vec{r}-\omega_k t+\theta)} + \hat{\varepsilon}^*_{k(L,R)} e^{-i(\vec{k}\cdot\vec{r}-\omega_k t+\theta)}]^2 d^3 r \right\rangle_T = \hbar\omega_k \tag{25}$$

where $\lambda = (L, R)$ correspond to left or right circular polarization, respectively.

We deduce the expression of the photon vector potential amplitude constant ξ

$$\xi = \pm \frac{1}{(2\pi)^{3/2}} \sqrt{\frac{\hbar}{8\alpha\varepsilon_0 c^3}} = \frac{\hbar}{4\pi e c} \tag{26}$$

where $\alpha = e^2/(4\pi\varepsilon_0)\hbar c \simeq 1/137$ is the fine structure constant and e is the electron–positron charge.

Introducing $\alpha_{0k}(\omega_k) = \xi\omega_k = (\hbar/4\pi e c)\omega_k$ in (1), we get the photon quantization volume issued now from the energy normalization:

$$V_k = \left(\frac{\hbar}{2\varepsilon_0 \xi^2}\right)\omega_k^{-3} = 4\alpha\lambda_k^3 \tag{27}$$

Note that (23), issued from the density of states, and (27), issued from the energy normalization, yield the same numerical result $V_k \simeq 0.029\ \lambda_k^3$, showing the self-consistency of the calculation.

Therein, it is important to note that the electron–positron charge, a fundamental physical constant, issues naturally during the energy normalization procedure and is directly related to the photon vector potential amplitude quantization constant ξ [40, 42, 43]:

$$e = (4\pi)^2 \alpha \frac{\xi}{\mu_0} \tag{28}$$

where μ_0 is the vacuum magnetic permeability.

The last expression relates the photon vector potential amplitude constant to the positive–negative elementary charge, which lays the basis for understanding the mechanisms of the mutual transformation photons $\leftrightarrow$ electrons + positrons as well as the vacuum polarization. It is worth investigating whether the electrons and positrons, more generally leptons and antileptons, have an internal structure composed of photons standing states in specific spatial configurations (toroidal or other), as already suggested previously [44].

3.2. *Single photon electric and magnetic fields*

We can now express the vector potential of an electromagnetic k-mode in classical and quantum representations using the defined vector potential amplitude in (24):

$$\begin{aligned}\vec{\alpha}_{k(L,R)}(\vec{r},t) &= \xi\omega_k\left[\hat{\varepsilon}_{k(L,R)}\, e^{i(\vec{k}\cdot\vec{r}-\omega_k t+\theta)} + \hat{\varepsilon}^*_{k(L,R)}\, e^{-i(\vec{k}\cdot\vec{r}-\omega_k t+\theta)}\right]\\ &= \omega_k\, \vec{\Xi}_{k(L,R)}(\vec{r},t) \qquad (29)\end{aligned}$$

$$\begin{aligned}\tilde{\alpha}_{k(L,R)}(\vec{r},t) &= \xi\omega_k\left[\hat{\varepsilon}_{k(L,R)}\, a_{k(L,R)} e^{i(\vec{k}\cdot\vec{r}-\omega_k t+\theta)}\right.\\ &\quad \left. + \hat{\varepsilon}^*_{k(L,R)}\, a^+_{k(L,R)} e^{-i(\vec{k}\cdot\vec{r}-\omega_k t+\theta)}\right] = \omega_k\, \tilde{\Xi}_{k(L,R)}(\vec{r},t)\end{aligned} \qquad (30)$$

The precession of the quantized vector potential around the propagation axis at the angular frequency ω_k generates orthogonal electric $\vec{\varepsilon}_k$ and magnetic $\vec{\beta}_k$ fields [25, 40, 45, 46], whose magnitudes, i.e., the square root of the sum of the squares of the components, are obtained straightforwardly:

$$\begin{aligned}\vec{\varepsilon}_k &= -\frac{\partial}{\partial t}\xi\omega_k\left(\hat{\varepsilon}_{k(L,R)} e^{i\left(\vec{k}\cdot\vec{r}-\omega_k t+\theta\right)} + \hat{\varepsilon}^*_{k(L,R)} e^{-i\left(\vec{k}\cdot\vec{r}-\omega_k t+\theta\right)}\right)\\ \rightarrow \|\vec{\varepsilon}_k\| &= \sqrt{2}|\xi|\omega_k^2 \qquad (31)\\ \vec{\beta}_k &= \vec{\nabla}\times\xi\omega_k\left(\hat{\varepsilon}_{k(L,R)} e^{i\left(\vec{k}\cdot\vec{r}-\omega_k t+\theta\right)} + \hat{\varepsilon}^*_{k(L,R)} e^{-i\left(\vec{k}\cdot\vec{r}-\omega_k t+\theta\right)}\right)\\ \rightarrow \|\vec{\beta}_k\| &= \sqrt{2\varepsilon_0\mu_0}|\xi|\omega_k^2 \qquad (32)\end{aligned}$$

Note that $\vec{\varepsilon}_k$ and $\vec{\beta}_k$ are proportional to the square of the frequency and independent of any external volume parameter. It is also worth noting that the proportionality of the photon electric field amplitude to the square of the frequency may naturally lead to a vacuum energy density compatible with recent astrophysical observations on the cosmic acceleration [47].

Experimental proposals have already been advanced previously [45] for validating the dependence of the photon electric field on the square of the frequency, which could also be an indirect validation for the vector potential dependence on frequency. As the linear Stark shift is directly proportional to the applied electric field, and the quadratic Stark shift to the square of the electric field, various configurations of quantum systems may be chosen for that purpose. Thus, high resolution spectroscopy might be used for the measurement of the Stark shifts induced by the mean electric field of photons upon the energy levels of selected quantum systems, atoms, ions or molecules.

Finally, the established photon electric and magnetic fields satisfy Maxwell's equations

$$\vec{\nabla} \times \vec{\varepsilon}_k = -\frac{\partial \vec{\beta}_k}{\partial t} \quad \vec{\nabla} \cdot \vec{\varepsilon}_k = 0 \tag{33}$$

$$\vec{\nabla} \times \vec{\beta}_k = \varepsilon_0 \mu_0 \frac{\partial \vec{\varepsilon}_k}{\partial t} \quad \vec{\nabla} \cdot \vec{\beta}_k = 0 \tag{34}$$

considered in vacuum and in the absence of charges.

3.3. *The photon quantization volume and Heisenberg's uncertainty*

The single photon vector potential amplitude operators are now written as

$$\tilde{\alpha}_{k\lambda} = \xi \omega_k \alpha_{k\lambda}; \quad \tilde{\alpha}^*_{k\lambda} = \xi \omega_k \alpha^+_{k\lambda} \tag{35}$$

On the other hand, the position and momentum Hermitian operators are expressed respectively as follows [1, 2, 18]:

$$\tilde{Q}_{k\lambda} = \sqrt{\varepsilon_0 V_k}\,(\tilde{\alpha}_{k\lambda} + \tilde{\alpha}^*_{k\lambda}); \quad \tilde{P}_{k\lambda} = -i\omega_k \sqrt{\varepsilon_0 V_k}\,(\tilde{\alpha}_{k\lambda} - \tilde{\alpha}^*_{k\lambda}) \tag{36}$$

Using now (27) and (35) in the last expressions of position and momentum operators (36), it is straightforward to demonstrate Heisenberg's commutation relation

$$\begin{aligned} [\tilde{Q}_{k\lambda}, \tilde{P}_{k'\lambda'}] &= -i\varepsilon_0 \omega^2_{k'} \omega_k \sqrt{V_k V_{k'}}\, \left[\left(\xi a_{k\lambda} + \xi\, a^+_{k\lambda} \right), \left(\xi a_{k'\lambda'} - \xi\, a^+_{k'\lambda'} \right) \right] \\ &= i\hbar \delta_{kk'} \delta_{\lambda\lambda'} \end{aligned} \tag{37}$$

showing that the position–momentum uncertainty for the photon, an intrinsic physical property, lies in the characteristic quantization volume.

Indeed, it is now easy to express the particle properties of the photon, energy and momentum, through the classical electromagnetic functions integrated over the quantization volume V_k and considering circular polarization:

$$E_k = \int_{V_k} 2\varepsilon_0 \alpha_{0k}^2(\omega_k)\omega_k^2 d^3r = \int_{V_k} 2\varepsilon_0 \xi^2 \omega_k^4 d^3r = 2\varepsilon_0 \xi^2 \omega_k^4 V_k = \hbar\omega_k \tag{38}$$

$$\vec{p}_k = \int_{V_k} \varepsilon_0 \vec{\varepsilon}_k \times \vec{\beta}_k \, d^3r = \varepsilon_0(\sqrt{2}\,\xi\omega_k^2)(\sqrt{2\varepsilon_0\mu_0}\xi\omega_k^2)V_k\vec{k}/|\vec{k}| = \hbar\vec{k} \tag{39}$$

where for the momentum we have used the photon electric and magnetic fields established above in (31) and (32).

Obviously, the photon energy and momentum are not carried by a point particle but by a wave–corpuscle with a precise quantization volume [25, 40]. Consequently, the photon is not a point or a singularity and constitutes a particular case in the standard particle model. The photon quantization volume also provides a physical explanation for why a photon position operator cannot be defined and can only be expressed through the quantization volume, as in the relation (36).

Finally, it is important to note that even though photons are not points, it has been demonstrated experimentally that they are indivisible [2, 18, 48], carrying energy and momentum, and consequently they behave exactly as Einstein had predicted [49].

3.4. *The photon wave function*

It can be easily shown that the vector potential function $\vec{\alpha}_{k(L,R)}(\vec{r},t)$ with the quantized amplitude expressed in (29) satisfies Maxwell's propagation equation in vacuum

$$\vec{\nabla}_k^2 \vec{\alpha}_{k(L,R)}(\vec{r},t) - \frac{1}{c^2}\frac{\partial^2}{\partial t^2}\vec{\alpha}_{k(L,R)}(\vec{r},t) = 0 \tag{40}$$

as well as Schrödinger's equation with the relativistic massless particle Hamiltonian

$$i\hbar\frac{\partial}{\partial t}\vec{\alpha}_{k(L,R)}(\vec{r},t) = \tilde{H}_k \vec{\alpha}_{k(L,R)}(\vec{r},t) \tag{41}$$

where $\tilde{H}_k = \hbar\tilde{\omega}_k = -i\,\hbar c\vec{\nabla}_k$, with $\tilde{\omega}_k$ the angular frequency operator $\tilde{\omega}_k = -i\,c\vec{\nabla}_k$, where $\vec{\nabla}_k$ is merely denoted to act upon the mode k.

Also, $\vec{\alpha}_{k(L,R)}(\vec{r},t)$ satisfies a Schrödinger-like equation for the vector potential amplitude

$$i\xi\frac{\partial}{\partial t}\vec{\alpha}_{k(L,R)}(\vec{r},t) = \tilde{\alpha}_{0k}\vec{\alpha}_{k(L,R)}(\vec{r},t) \tag{42}$$

with the vector potential amplitude operator, which is also written as $\tilde{\alpha}_{0k} = \xi\tilde{\omega}_k = -i\,\xi c\vec{\nabla}_k$.

The last two symmetrical equations can be coupled together, leading to the photon *vector potential–energy* equation, a unique quantum equation expressing both the first (energy) and second (vector potential) quantization of the electromagnetic field [25, 40, 46]:

$$i\begin{pmatrix}\xi\\ \hbar\end{pmatrix}\frac{\partial}{\partial t}\vec{\alpha}_{k(L,R)}(\vec{r},t) = \begin{pmatrix}\tilde{\alpha}_{0k}\\ \tilde{H}_k\end{pmatrix}\vec{\alpha}_{k(L,R)}(\vec{r},t) \tag{43}$$

The constants ξ and $\hbar$ characterize the quantization of the photon vector potential amplitude $\alpha_{0k}(\omega_k) = \xi\omega_k$ and energy $E_k(\omega_k) = \hbar\omega_k$, respectively.

A photon wave function related to the energy density of a single photon has already been established previously, based also on the vector potential function [41]. Now, we will establish a photon wave function as a probability amplitude according to the quantum mechanics principles. Indeed, in quantum theory, the wave function normalization is generally carried out in such a way that the integration of the square modulus of the wave function within a given volume expresses the localization probability in the said volume.

According to these concepts, as the experiments have shown that a single photon is localizable only within a volume proportional to λ_k^3, we may establish a wave function for the photon based precisely on the vector potential function $\vec{\alpha}_{k(L,R)}(\vec{r},t)$ and by taking into account the photon quantization volume V_k for the normalization. Consequently, the photon wave function can be defined as follows [50, 51]:

$$\vec{\Phi}_{k(L,R)}(\vec{r},t) = \left(\frac{\varepsilon_0\omega_k}{\hbar}\right)^{1/2}\left[\alpha_{0k}(\omega_k)(\hat{\varepsilon}_{k(L,R)}e^{i(\vec{k}\cdot\vec{r}-\omega_k t+\theta)} + \hat{\varepsilon}^*_{k(L,R)}e^{-i(\vec{k}\cdot\vec{r}-\omega_k t+\theta)})\right] \tag{44}$$

where the photon spin is expressed through the helicity, with left or right circular polarizations corresponding to the values $\pm\hbar$.

The normalization pre-factor has been chosen in such a way that

$$\begin{aligned}|\vec{\Phi}_{k(L,R)}(\vec{r},t)|^2 &= \frac{\varepsilon_0\omega_k}{\hbar}\left[\xi^2\omega_k^2\left(\left|\hat{\varepsilon}_{k(L,R)}\right|^2+\left|\hat{\varepsilon}^*_{k(L,R)}\right|^2\right)\right]\\ &= \frac{2\varepsilon_0\xi^2\omega_k^3}{\hbar} = \frac{1}{V_k}\end{aligned} \tag{45}$$

entailing that the normalization condition is satisfied with respect to the quantization volume

$$\int_{V_k}\left|\vec{\Phi}_{k(L,R)}(\vec{r},t)\right|^2 d^3r' = 1 \tag{46}$$

Therefore, when a photon is emitted in vacuum at an instant t_0 at the coordinate r_0, the probability to be localized at t around $r = r_0 + ct$ on the propagation axis is normalized with respect to the quantization volume V_k. Furthermore, $\vec{\Phi}_{k(L,R)}(\vec{r},t)$ satisfies Maxwell's propagation equation in vacuum

$$\vec{\nabla}_k^2\vec{\Phi}_{k(L,R)}(\vec{r},t) - \frac{1}{c^2}\frac{\partial^2}{\partial t^2}\vec{\Phi}_{k(L,R)}(\vec{r},t) = 0 \tag{47}$$

as well as the photon *vector potential–energy* quantum equation

$$i\begin{pmatrix}\xi\\ \hbar\end{pmatrix}\frac{\partial}{\partial t}\vec{\Phi}_{k(L,R)}(\vec{r},t) = \begin{pmatrix}\tilde{\alpha}_{0k}\\ \tilde{H}_k\end{pmatrix}\vec{\Phi}_{k(L,R)}(\vec{r},t) \tag{48}$$

Consequently, $\vec{\Phi}_{k(L,R)}(\vec{r},t)$ is a real wave function for the photon involving the spin through the helicity corresponding to circular polarizations and normalized with respect to the quantization volume V_k.

Obviously, the photon wave function $\vec{\Phi}_{k(L,R)}(\vec{r},t)$ is defined at an instant t at the coordinate r. However, the photon as an integral wave-corpuscle may only be conceived within a volume V_k around r, as expressed in (46). In addition, its physical properties are obtained by integration of the wave properties over V_k, according to (38) and (39). Consequently, in the case of the photon, the mean values of the energy and momentum operators can be calculated by considering the quantization volume V_k. Hence, using the established photon wave function, the mean value of the relativistic massless particle Hamiltonian in a single photon state is

$$
\begin{aligned}
&|\langle\Phi_{k(L,R)}(\vec{r},t)|\tilde{H}_k|\Phi_{k(L,R)}(\vec{r},t)\rangle_{V_k}| \\
&= |\langle\Phi_{k(L,R)}(\vec{r},t)| - i\hbar c\vec{\nabla}_k|\Phi_{k(L,R)}(\vec{r},t)\rangle_{V_k}| \\
&= \varepsilon_0\xi^2\omega_k^4 V_k|(e^{i(\vec{k}\cdot\vec{r}-\omega_k t+\theta)}\hat{\varepsilon}_{k(L,R)} + e^{-i(\vec{k}\cdot\vec{r}-\omega_k t+\theta)}\hat{\varepsilon}^*_{k(L,R)}) \\
&\quad \times (e^{i(\vec{k}\cdot\vec{r}-\omega_k t+\theta)}\hat{\varepsilon}_{k(L,R)} - e^{-i(\vec{k}\cdot\vec{r}-\omega_k t+\theta)}\hat{\varepsilon}^*_{k(L,R)})| \\
&= \varepsilon_0\xi^2\omega_k^4 V_k|(e^{2i(\vec{k}\cdot\vec{r}-\omega_k t+\theta)} - e^{-2i(\vec{k}\cdot\vec{r}-\omega_k t+\theta)})| \\
&= (2\varepsilon_0\xi^2\omega_k^4)V_k|\sin 2(\vec{k}\cdot\vec{r}-\omega_k t+\theta)| \qquad (49)
\end{aligned}
$$

The phase function $|\sin 2(\vec{k}\cdot\vec{r}-\omega_k t+\theta)|$ varies from zero to one, which physically means that when $|\sin 2(\vec{k}\cdot\vec{r}-\omega_k t)+\theta| = 1$ corresponds to the total space of the quantization volume V_k, so that the mean value of the Hamiltonian gives precisely Planck's single photon energy $|\langle\Phi_{k(L,R)}(\vec{r},t)|\tilde{H}_k|\Phi_{k(L,R)}(\vec{r},t)\rangle_{V_k}| = \hbar\omega_k$.

With the same token, the momentum operator yields Einstein's single photon momentum

$$
\begin{aligned}
&|\langle\Phi_{k(L,R)}(\vec{r},t)|\vec{p}_k|\Phi_{k(L,R)}(\vec{r},t)\rangle_{V_k}| \\
&= |\langle\Phi_{k(L,R)}(\vec{r},t)| - i\hbar\vec{\nabla}_k|\Phi_{k(L,R)}(\vec{r},t)\rangle_{V_k}| = \frac{\hbar\omega_k}{c} \qquad (50)
\end{aligned}
$$

It is also straightforward to calculate the energy density for a single photon state

$$
\begin{aligned}
W_k &= |\Phi_{k(L,R)}(\vec{r},t)|^2|\langle\Phi_{k(L,R)}(\vec{r},t)|\tilde{H}_k|\Phi_{k(L,R)}(\vec{r},t)\rangle_{V_k}| \\
&= \frac{\hbar\omega_k}{V_k} = 2\varepsilon_0\xi^2\omega_k^4 \qquad (51)
\end{aligned}
$$

which, by considering the photon electric and magnetic fields established in (31) and (32), is exactly equal to the classical expression

$$
W_k = \frac{1}{2}\left(\varepsilon_0\left|\vec{\varepsilon}_k\right|^2 + \frac{1}{\mu_0}\left|\vec{\beta}_k\right|^2\right) = 2\varepsilon_0\xi^2\omega_k^4 \qquad (52)
$$

We have thus demonstrated the perfect equivalence between the classical and quantum energy densities for a single photon, reappraising the relation (1).

Note also that the obtained single photon energy density is proportional to the fourth power of the frequency, as in the case of the radiation emitted by a classical oscillating dipole [19, 20].

Furthermore, a Poynting vector can now be readily defined for a single k-mode photon:

$$\vec{S}_k = \frac{1}{\mu_0}\vec{\varepsilon}_k \times \vec{\beta}_k = W_k\, c\frac{\vec{k}}{|\vec{k}|} \tag{53}$$

To conclude, the photon wave function $\vec{\Phi}_{k(L,R)}(\vec{r}, t)$ expressed in (44) represents a probability density normalized with respect to the intrinsic photon quantization volume V_k. In addition, it satisfies Maxwell's propagation equation in vacuum, Schrödinger's equation with the relativistic massless particle Hamiltonian and a Schrödinger-like equation for the vector potential amplitude [50].

Excellent studies have been published previously on the intriguing photon wave function [12–17]. However, none of those considers the quantization of the vector potential amplitude at a single photon state, although it has been demonstrated experimentally that the photon, in the sense of Einstein's concept, is a local indivisible quantum of the electromagnetic field. Consequently, the energy quantization should imply physically the single photon vector potential quantization.

From the experimental point of view, the investigation of a cavity-free photon wave function should not be an easy task. Nonetheless, the measurement of the minimum quantization volume, used for the wave function spatial normalization, would be of great importance. Thus, the minimization of microwave cavities having homothetic dimensions with the experimental values [32–35] for the single-mode wavelength (λ_k) and lateral extension ($\sim\lambda_k/4$) might be useful.

In the IR frequency range, the optimization of the shorter photoluminescence lifetime of a quantum dot by using various dimensions of micro cavities [51, 52] might also be used for that purpose.

3.5. *The electromagnetic vacuum*

According to (24), (31), (32) and (38), for $\omega_k = 0$, the photon vector potential, electric and magnetic fields as well as the energy vanish. This physical state lies beyond the Bohm–Aharonov effect, in which the electric and magnetic fields are absent, but a vector potential subsists in space. In fact, a photon, as an integral physical entity, can only be conceived for a non-zero frequency which corresponds to the precession of the vector potential amplitude with circular polarization over a wavelength along the propagation axis. However, the zero-frequency state does not correspond to an intangible empty space because in the absence of precession, the vector potential amplitude per frequency still subsists. In fact, the principal function $\Xi_{k(L,R)}$ of the vector potential as expressed in (29) and (30) does not cancel and is written in both classical and quantum representations as

$$\vec{\Xi}_{0(L,R)} = \xi(\hat{\varepsilon}_{(L,R)}\, e^{i\theta} + \hat{\varepsilon}^*_{(L,R)}\, e^{-i\theta}) \tag{54}$$

$$\tilde{\Xi}_{0(L,R)} = \xi(\hat{\varepsilon}_{(L,R)}\, a_{k(L,R)} e^{i\theta} + \hat{\varepsilon}^*_{(L,R)}\, a^+_{k(L,R)} e^{-i\theta}) \tag{55}$$

Hence, in the absence of energy, vector potential as well as electric and magnetic fields, the zero-frequency state $\Xi_{0(L,R)}$ represents the electromagnetic field ground state, which corresponds to the electromagnetic vacuum. Obviously, $\Xi_{0(L,R)}$ is a real field permeating all of space, involving the circular polarization information and having an electric potential nature with positive and negative components. It is worth noting that it has both classical and quantum representations, entailing the introduction of the new fundamental concept that the electromagnetic vacuum can also be defined in the classical theory.

It is clear that the phase parameter θ can take any values, and consequently, the electromagnetic vacuum is composed of all possible states $\Xi_{k(L,R)}(\vec{r}, t)$:

$$\vec{\Xi}_{k(L,R)}(\vec{r}, t) = \xi(\hat{\varepsilon}_{k(L,R)}\, e^{i(\vec{k}\cdot\vec{r}-\omega_k t+\phi)} + \hat{\varepsilon}^*_{k(L,R)}\, e^{-i(\vec{k}\cdot\vec{r}-\omega_k t+\phi)}) \tag{56}$$

$$\begin{aligned} &\tilde{\Xi}_{k(L,R)}(\vec{r}, t) \\ &\quad = \xi(\hat{\varepsilon}_{k(L,R)}\, a_{k(R,L)} e^{i(\vec{k}\cdot\vec{r}-\omega_k t+\phi)} + \hat{\varepsilon}^*_{k(L,R)} a^+_{k(L,R)} e^{-i(\vec{k}\cdot\vec{r}-\omega_k t+\phi)}) \end{aligned} \tag{57}$$

From the photon *vector potential–energy* equation, we get the equation governing the vacuum states

$$i\,\frac{\partial}{\partial t}\begin{bmatrix} \vec{\Xi}_{k(L,R)}(\vec{r}, t) \\ \tilde{\Xi}_{k(L,R)}(\vec{r}, t) \end{bmatrix} = \tilde{\omega}_k \begin{bmatrix} \vec{\Xi}_{k(L,R)}(\vec{r}, t) \\ \tilde{\Xi}_{k(L,R)}(\vec{r}, t) \end{bmatrix} = \begin{bmatrix} \vec{\alpha}_{k(L,R)}(\vec{r}, t) \\ \tilde{\alpha}_{k(L,R)}(\vec{r}, t) \end{bmatrix} \tag{58}$$

The physical meaning of Eq. (58) is that real photons emerge from the vacuum states under the action of the angular frequency operator $\tilde{\omega}_k = -i\,c\vec{\nabla}_k$.

Consequently, using the electromagnetic vacuum, we can now demonstrate the calculation of the spontaneous emission probability, lifting the well-known QED ambiguity related to the electron–vacuum interaction Hamiltonian. We recall that the Hamiltonian of an atom in the presence of an electromagnetic field is written in QED formalism within the dipole approximation as [1,2]

$$\begin{aligned} H_{\text{tot}} &= H_{\text{atom}} + H_{EM} + H_{\text{int}} \\ &= \sum_i \hbar\omega_i |\Psi_i\rangle\langle\Psi_i| + \sum_{k,\lambda} \hbar\omega_k \left(a_{k\lambda}^{+} a_{k\lambda} + \frac{1}{2}\right) - \vec{D}\cdot\vec{E}(\vec{r},t) \end{aligned} \tag{59}$$

where $|\Psi_i\rangle$ in H_{atom} represent the atomic quantum levels with respective energies $\hbar\omega_i$, H_{EM} is the electromagnetic field harmonic oscillator Hamiltonian including the zero-point energy $\sum_{k,\lambda} \frac{1}{2}\hbar\omega_k$, $\vec{D} = e\vec{r}$ is the dipole moment of the bound electron and $\vec{E}(\vec{r},t)$ is the radiation electric field involved in the interaction Hamiltonian H_{int}.

The electromagnetic radiation electric field $\vec{E}(\vec{r},t)$ is calculated in standard QED from Heisenberg's equation of motion

$$\vec{E}(\vec{r},t) = -\frac{\partial\vec{A}(\vec{r},t)}{\partial t} = \frac{i}{\hbar}\left[\vec{A}(\vec{r},t), \sum_{k,\lambda} \hbar\omega_k \left(a_{k\lambda}^{+} a_{k\lambda} + \frac{1}{2}\right)\right] \tag{60}$$

As generally quoted in the literature, the spontaneous emission rate in QED is calculated, *neglecting* the zero-point energy term in the electromagnetic field Hamiltonian [1,2]. Furthermore, the zero-point energy is a constant, so that $[\vec{A}(\vec{r},t), \sum_{k,\lambda} \frac{1}{2}\hbar\omega_k] = 0$ in the radiation electric field $\vec{E}(\vec{r},t)$ calculated by (60), and consequently, the vacuum term has absolutely no contribution to the interaction Hamiltonian used to calculate the spontaneous emission in QED.

Surprisingly, with the aim of calculating the interaction between an electron and the quantum vacuum, the zero-point energy is cast aside in both Hamiltonians H_{EM} and H_{int}.

As mentioned previously, this well-known ambiguity in QED is due to the fact that the spontaneous emission is calculated based on the properties of the photon creation and annihilation operators,

$a_{k\lambda}^{+}$ and $a_{k\lambda}$, respectively, and not by using the zero-point energy. In addition, it comes out from (60) that, since the vacuum state in QED is not described by a function of $a_{k\lambda}^{+}$ and $a_{k\lambda}$ operators, an electron–vacuum interaction Hamiltonian cannot be defined.

Conversely, the electromagnetic vacuum states expressed by $\tilde{\Xi}_{k(L,R)}(\vec{r},t)$ in (57) do depend on the single photon creation and annihilation operators. Consequently, it is straightforward to define an interaction Hamiltonian per mode between an electron and the vacuum:

$$H_{\text{int}} = \frac{\hbar e}{m_e c}\tilde{\Xi}_{k(L,R)}\tilde{\omega}_k = -i\hbar\frac{e}{m_e}\tilde{\Xi}_{k(L,R)}\vec{\nabla}_k \tag{61}$$

where m_e is the electron mass.

Using (61) and applying first-order time-dependent perturbation theory, it has been shown that the spontaneous emission rate is readily calculated with a similar procedure as in the QED theory [39,40]. The Lamb shift is calculated as usual using the photon creation operator, while it can be readily demonstrated that any particle accelerated in the electromagnetic vacuum field experiences the Unruh temperature [43]. Finally, based on the fact that the single photon electric field amplitude is proportional to the square of the frequency, it can be shown that the electromagnetic vacuum fluctuations reproduce a vacuum energy density comparable to the astrophysical observations [47]. Following these developments, the pic of the electromagnetic vacuum fluctuations should be located in the THz frequencies, and therefore, a detailed investigation of the James Webb Space Telescope could give an answer to that issue.

Consequently, the electromagnetic vacuum complements the normal ordering Hamiltonian and confers a real physical state to the vacuum expressed in both classical and quantum theories.

As discussed above, a k-mode photon is generated from the application of the angular frequency ω_k upon the corresponding vacuum k-mode state $\Xi_{k(L,R)}(\vec{r},t)$. Thus, it becomes clear that the electromagnetic vacuum is the real physical basis of the photon and of its wave function:

$$\vec{\Phi}_{k(L,R)}(\vec{r},t) = \left(\frac{\varepsilon_0\omega_k^3}{\hbar}\right)^{1/2}\vec{\Xi}_{k(L,R)}(\vec{r},t) \tag{62}$$

This leads to the interesting conclusion that when a k-mode photon with polarization λ is created in space, the electromagnetic vacuum is depleted by the corresponding k-mode and λ-polarization vacuum state. We may consider that immediately after this process, the new photon created at the same frequency carries an opposite polarization, laying down new possibilities for understanding entanglement. The process does not depend on the distance separating the photon pair but the probability to occur should diminish at high temperatures.

4. Conclusion

The vector potential quantization at a single photon level remedies the QED fundamental drawbacks. Thus, the photon vector potential amplitude operators as well as the electric and magnetic field amplitudes are defined independently of any external volume parameter. Also, the single photon electric and magnetic field amplitudes in vacuum are directly proportional to the square of the frequency.

Regarding the principal singularity related to the zero-point energy, it can be eliminated by considering the normal ordering Hamiltonian associated with the electromagnetic vacuum, which has both classical and quantum representations. The last one is a zero-energy universal field expressed through the photon creation and annihilation operators and permits us to describe the interaction between the electrons and the vacuum states. This removes the ambiguity related to the zero-point energy in the interaction Hamiltonian for the vacuum effects without compromising any of the achievements in QED. The concept of the harmonic oscillator Hamiltonian for the electromagnetic field, resulting from the unjustified mean value of the normal and anti-normal order Hamiltonians, has to be appraised.

In addition, the density of states theory ascribes an intrinsic quantization volume to a single photon. The energy normalization of a single-mode plane electromagnetic wave over a period confirms precisely the photon quantization volume issued from the density of states theory. Hence, the quantum properties of a single photon, energy and momentum, are not carried by a point but by a wave–corpuscle characterized by the quantization volume whose spatial dimensions are proportional to the cube of the wavelength, as confirmed experimentally. Heisenberg's uncertainty of the photon position and momentum is due precisely to its intrinsic quantization volume.

Consequently, single photons are neither point particles nor harmonic oscillators, and they constitute a particular case in the standard particle model.

A photon wave function is established, normalized with respect to the quantization volume, satisfying Schrödinger's equation for the energy and a Schrödinger-like equation for the vector potential as well as Maxwell's propagation equation in vacuum. Furthermore, it depends directly on the electromagnetic vacuum states, envisioning new horizons for understanding entanglement. It is drawn that photons are free states of the electromagnetic vacuum while elementary charges may be local standing states of the same vacuum field. The established formalism and equation (28) show that leptons-antileptons can be created from photons and inversely while the equation (59), governing the vacuum states, entails that under specific experimental conditions photons can be created from the vacuum.

Finally, we have advanced some ideas for carrying out specific experiments in order to investigate the frequency dependence of the photon electric field and vector potential as well as for the definition of the intrinsic photon volume for the wave function normalization. Certainly, these questions are open, and any other experimental proposals are welcome toward that issue.

References

[1] M. Weissbluth, *Photon-Atom Interactions* (Academic Press, Cambridge, MA, USA, 1988).

[2] J. C. Garrison and R. Y. Chiao, *Quantum Optics* (Oxford University Press, New York, NY, USA, 2008).

[3] G. Auletta, *Foundations and Interpretation of Quantum Mechanics* (World Scientific, Singapore, 2001).

[4] A. G. Riess *et al.*, Observational Evidence from Supernovae for an Accelerating Universe and a Cosmological Constant, *Astron. J.* 116, 1009 (1998).

[5] S. Perlmutter *et al.*, Measurements of Ω and Λ from 42 High-Redshift Supernovae, *Astrophys. J.* 517, 565 (1999).

[6] A. G. Riess *et al.*, Type Ia Supernova Discoveries at z >1 from the Hubble space Telescope: Evidence for past deceleration and constraints on dark energy evolution, *Astrophys. J.* 607, 665 (2004).
[7] G. R. Bengochea, G. León, E. Okon, and D. Sudarsky, Can the quantum vacuum fluctuations really solve the cosmological constant problem?, *Eur. Phys. J. C* 80, 18 (2020).
[8] D. Scolnic *et al.*, Supercal: Cross-calibration of multiple photometric systems to improve cosmological measurements with type Ia Supernovae, *Astrophys. J.* 815, 117 (2015).
[9] T. Padmanabhan, Cosmological constant - the weight of the vacuum, *Phys. Rep.* 380, 235 (2003).
[10] J. A. Frieman, M. S. Turner, and D. Huterer, Dark energy and the accelerating Universe, *Annu. Rev. Astron. Astrophys.* 46, 385 (2008).
[11] B. H. Bransden and C. J. Joachain, *Physics of Atoms and Molecules*, 2nd ed. (Longman, New York, NY, USA, 2003).
[12] J. E. Sipe, Photon wave functions, *Phys. Rev. A* 52, 1875 (1995).
[13] I. Bialynicki-Birula, Photon wave function, in *Progress in Optics*, Vol. XXXVI, E. Wolf, Ed. (Elsevier, Amsterdam, The Netherlands, 1996).
[14] D. H. Kobe, A relativistic Schrödinger-like equation for photon and its second quantization, *Found. Phys.* 29, 1203 (1999).
[15] B. J. Smith and M. G. Raymer, Photon wave functions, wave packet quantization of light and coherence theory, *New J. Phys.* 9, 414 (2007).
[16] V. Debierre, The Photon Wave Function in Principle and in Practice, Ph.D. thesis, Ecole Centrale Marseille, Marseille, France, 2015.
[17] J. Cugnon, The Photon Wave Function, *Open J. Microphys.* 1, 41 (2011).
[18] G. Grynberg, A. Aspect, and C. Fabre, *Quantum Optics* (Cambridge University Press, New York, NY, USA, 2010).
[19] J. D. Jackson, *Classical Electrodynamics* (Wiley, Hoboken, NJ, USA, 1998).
[20] F. H. Read, *Electromagnetic radiation* (John Wiley & Sons, New York, NY, USA, 1980).
[21] T. A. Garrity, *Electricity and Magnetism for Mathematicians* (Cambridge University Press, New York, NY, USA, 2015).
[22] T. K. Das, *Quantum Mechanics, Axiomatic Approach and Understanding Through Mathematics* (Springer Nature, Singapore, 2023).
[23] H. Ibach and H. Lüth, *Solid-State Physics* (Springer-Verlag, Berlin, Germany, 1996).
[24] C. Meis, *Light and Vacuum*, 1st ed. (World Scientific, Singapore, 2014).

[25] C. Meis, Quantized Field of Single Photons, in *Single Photon Manipulation*, K. Xia, Ed. (IntechOpen, London, UK, 2019), pp. 15–36.

[26] P. W. Milonni, *The Quantum Vacuum* (Academic Press, London, UK, 1994).

[27] J. Schwinger, L. L. DeRaad Jr., and K. A. Milton, Casimir effect in dielectrics, *Ann. Phys.* 115, 1 (1978).

[28] P. W. Milonni, Casimir forces without the vacuum radiation, *Phys. Rev. A* 25, 1315 (1982).

[29] H. Ezawa, K. Nakamura, and K. Watanabe, The Casimir Force from Lorentz's, in *Frontiers in Quantum Physics*, S. C. Lim, R. Abd-Shukor, and K. H. Kwek, Eds. (Springer, Singapore, 1998), pp. 160–169.

[30] G. Gründler, The Casimir effect: No manifestation of zero-point energy, arXiv:1303.3790 (2013).

[31] A. I. Akhiezer and V. B. Berestetskii, *Quantum Electrodynamics* (Interscience, New York, NY, USA, 1965).

[32] H. L. Robinson, Diffraction patterns in circular apertures less than one wavelength in diameter, *J. Appl. Phys.* 24, 35 (1953).

[33] R. K. Hadlock, Diffraction patterns at the plane of a slit in a reflecting screen, *J. Appl. Phys.* 29, 918 (1958).

[34] G. Hunter and R. L. P. Wadlinger, Photons and neutrinos as electromagnetic solitons, *Phys. Essays* 2, 158 (1989).

[35] G. Hunter, in *Quantum uncertainties*, NATO ASI Series B, Vol. 162 (1987), p. 331.

[36] L. Mandel, Configuration-space photon number operators in quantum optics, *Phys. Rev.* 144, 1071 (1966).

[37] R. L. Pfleegor and L. Mandel, Interference of independent photon beams, *Phys. Rev.* 159, 1084 (1967).

[38] C. Meis, Photon wave-particle duality and the virtual electromagnetic waves, *Found. Phys.* 27, 865 (1997).

[39] C. Meis, Vector potential quantization and the quantum vacuum, *Phys. Res. Int.* 2014, 187432 (2014).

[40] C. Meis, *Light and Vacuum*, 2nd ed. (World Scientific, Singapore, 2017).

[41] C. Meis and P. R. Dahoo, Quantized vector potential and the photon wave-function, *J. Phys.: Conf. Ser.* 936, 012004 (2017).

[42] C. Meis, The electromagnetic field ground state and the cosmological evolution, *J. Phys.: Conf. Ser.* 1141, 012072 (2018).

[43] C. Meis, Primary role of the quantum electromagnetic vacuum in Gravitation and Cosmology, in *Cosmology 2020 — The Current State*, M. L. Smith, Ed. (IntechOpen, London, UK, 2020).

[44] J. G. Williamson and M. B. van der Mark, Is the electron a photon with toroidal topology?, *Ann. Fond. Louis de Broglie* 22, 133 (1997).

[45] C. Meis and P. R. Dahoo, Vector potential quantization and the photon intrinsic electromagnetic properties: Towards nondestructive photon detection, *Int. J. Quantum Inf.* 15, 1740003 (2017).

[46] C. Meis, Electromagnetic wave representation as a superposition of the quantized vector potential field of single photons, *Int. J. Appl. Electromagn. Mech.* 71, S57 (2023).

[47] C. Meis, The vacuum electromagnetic energy density - Towards a physical comprehension of the vacuum energy scale problem, *Int. J. Mod. Phys. A* 35, 2050153 (2020).

[48] P. Grangier, G. Roger, and A. Aspect, Experimental evidence for a photon anticorrelation effect on a beam splitter: A new light on single-photon interferences, *Europhys. Lett.* 1, 173 (1986).

[49] A. Einstein, *The Collected Papers of Albert Einstein*, J. Stachel, D. C. Cassidy, J. Renn, and R. Schulmann, Eds. (Princeton University Press, Princeton, NJ, USA, 1987).

[50] C. Meis, Photon Structure and Wave Function from the Vector Potential Quantization, *J. Mod. Phys.* 14, 311 (2023).

[51] G. M. Gérard and B. Gayral, Strong Purcell effect from InAs quantum boxes in three-dimensional solid-state microcavities, *J. Lightwave Technol.* 17, 2089 (1999).

[52] S. L. Chuang, *Physics of Photonic Devices*, 2nd ed. (Wiley, Hoboken, NJ, USA, 2009).

Conclusion

In Part I, we have investigated the vector potential quantization enhanced at a single photon level for a k-mode photon with frequency ω_k and wavelength λ_k. Here, the results pertaining to the themes covered in this part are summarized, listing the following new insights that have been established:

- Photon vector potential quantization independent of any external cavity parameter

$$\alpha_{0k}(\omega_k) = \xi \omega_k$$

- Quantization constant of the photon vector potential amplitude

$$\xi = \frac{1}{(2\pi)^{3/2}} \sqrt{\frac{\hbar}{8\alpha\varepsilon_0 c^3}} = \frac{\hbar}{4\pi e c} = \pm 1.747\ 10^{-25}\ \text{Volt m}^{-1}\text{s}^2$$

- Photon vector potential amplitude operator independent of any external cavity

$$\text{Wave: } \tilde{\alpha}_{0k} = -i\xi c \vec{\nabla}_k;$$

$$\text{Particle: } \tilde{\alpha}_{0k\lambda} = \xi \omega_k a_{k\lambda} \quad \text{and} \quad \tilde{\alpha}^{+}_{0k\lambda} = \xi \omega_k a^{+}_{k\lambda}$$

- Relation between the electron–positron charge and the photon vector potential

$$e = (4\pi)^2 \alpha \xi / \mu_0$$

  where α is the fine structure constant and μ_0 the vacuum magnetic permeability.

- Electromagnetic vacuum representation in both classical and quantum electromagnetic theory

$$\vec{\Xi}_{k(L,R)}(\vec{r},t) = \xi\left(\hat{\varepsilon}_{k(L,R)}\, e^{i(\vec{k}\cdot\vec{r}-\omega_k t+\phi)} + \hat{\varepsilon}^*_{k(L,R)}\, e^{-i(\vec{k}\cdot\vec{r}-\omega_k t+\phi)}\right)$$

$$\tilde{\Xi}_{k(L,R)}(\vec{r},t) = \xi\left(\hat{\varepsilon}_{k(L,R)}\, a_{k(R,L)} e^{i(\vec{k}\cdot\vec{r}-\omega_k t+\phi)} \right.$$

$$\left. + \hat{\varepsilon}^*_{k(L,R)} a^+_{k(L,R)} e^{-i(\vec{k}\cdot\vec{r}-\omega_k t+\phi)}\right)$$

- Relation of photons, electrons and positrons to the electromagnetic vacuum: Photons as well as electrons and positrons issue from the electromagnetic vacuum field, and they are strongly related through the vacuum constant ξ.

- Photon electric and magnetic field magnitudes independent of any external volume parameter

$$\vec{\varepsilon}_k = -\frac{\partial}{\partial t}\xi\omega_k\left(\hat{\varepsilon}_{k(L,R)} e^{i(\vec{k}\cdot\vec{r}-\omega_k t+\theta)} + \hat{\varepsilon}^*_{k(L,R)} e^{-i(\vec{k}\cdot\vec{r}-\omega_k t+\theta)}\right)$$

$$\rightarrow \|\vec{\varepsilon}_k\| = \sqrt{2}|\xi|\omega_k^2$$

$$\vec{\beta}_k = \vec{\nabla}\times\xi\omega_k\left(\hat{\varepsilon}_{k(L,R)} e^{i(\vec{k}\cdot\vec{r}-\omega_k t+\theta)} + \hat{\varepsilon}^*_{k(L,R)} e^{-i(\vec{k}\cdot\vec{r}-\omega_k t+\theta)}\right)$$

$$\rightarrow \|\vec{\beta}_k\| = \sqrt{2\varepsilon_0\mu_0}|\xi|\omega_k^2$$

- Photon quantization volume

$$V_k = \left(\frac{\hbar}{2\varepsilon_0\xi^2}\right)\omega_k^{-3} = 4\alpha\lambda_k^3$$

where α is the fine structure constant.

- Photon quantum equation representing both the first (energy) and second (vector potential) quantization of the electromagnetic field

$$i\begin{pmatrix}\xi\\ \hbar\end{pmatrix}\frac{\partial}{\partial t}\vec{\alpha}_{k(L,R)}(\vec{r},t) = \begin{pmatrix}\tilde{\alpha}_{0k}\\ \tilde{H}_k\end{pmatrix}\vec{\alpha}_{k(L,R)}(\vec{r},t)$$

Eigenvalues are the photon vector potential amplitude and energy:

$$\alpha_{0k}(\omega_k) = \xi\omega_k; \quad E_k(\omega_k) = \hbar\omega_k$$

- Photon wave function as a probability density with respect to the quantization volume V_k

$$\vec{\Phi}_{k(L,R)}(\vec{r},t) = \left(\frac{\varepsilon_0 \omega_k}{\hbar}\right)^{1/2} \Big[\alpha_{0k}(\omega_k)\Big(\hat{\varepsilon}_{k(L,R)} e^{i(\vec{k}\cdot\vec{r}-\omega_k t+\theta)} + \hat{\varepsilon}^*_{k(L,R)} e^{-i(\vec{k}\cdot\vec{r}-\omega_k t+\theta)}\Big)\Big]$$

satisfies Maxwell's wave propagation equation

$$\vec{\nabla}^2_k \vec{\Phi}_{k(L,R)}(\vec{r},t) - \frac{1}{c^2}\frac{\partial^2}{\partial t^2}\vec{\Phi}_{k(L,R)}(\vec{r},t) = 0$$

Helmholtz equation

$$\vec{\nabla}^2_k \vec{\Phi}_{k(L,R)}(\vec{r},t) + k^2 \vec{\Phi}_{k(L,R)}(\vec{r},t) = 0$$

as well as the photon quantum equation, expressing the first and second quantization of the electromagnetic field:

$$i\begin{pmatrix}\xi\\ \hbar\end{pmatrix}\frac{\partial}{\partial t}\vec{\Phi}_{k(L,R)}(\vec{r},t) = \begin{pmatrix}\tilde{\alpha}_{0k}\\ \tilde{H}_k\end{pmatrix}\vec{\Phi}_{k(L,R)}(\vec{r},t)$$

where $\vec{\nabla}_k$ merely signifies that the gradient acts upon the mode k.

- Photon wave function normalization with respect to the quantization volume V_k

$$\int_{V_k} |\vec{\Phi}_{k(L,R)}(\vec{r},t)|^2 d^3r' = 1$$

- Relation between the photon wave function and the electromagnetic vacuum states

$$\vec{\Phi}_{k(L,R)}(\vec{r},t) = \left(\frac{\varepsilon_0 \omega_k^3}{\hbar}\right)^{1/2} \vec{\Xi}_{k(L,R)}(\vec{r},t)$$

- Photons/electromagnetic waves are oscillations of the electromagnetic vacuum states

$$i\frac{\partial}{\partial t}\begin{bmatrix}\vec{\Xi}_{k(L,R)}(\vec{r},t)\\ \tilde{\Xi}_{k(L,R)}(\vec{r},t)\end{bmatrix} = \tilde{\omega}_k \begin{bmatrix}\vec{\Xi}_{k(L,R)}(\vec{r},t)\\ \tilde{\Xi}_{k(L,R)}(\vec{r},t)\end{bmatrix} = \begin{bmatrix}\vec{\alpha}_{k(L,R)}(\vec{r},t)\\ \tilde{\alpha}_{k(L,R)}(\vec{r},t)\end{bmatrix}$$

where the angular frequency operator writes $\tilde{\omega}_k = -i\, c\vec{\nabla}_k$.

Part II

From Electromagnetic Vacuum to Cosmology

Introduction

In this part, the following new concepts are investigated:

- photons, electrons, positrons and electromagnetic vacuum;
- electromagnetic vacuum energy density and dark energy;
- cosmological constant;
- mass-charge equivalence;
- Coulomb's electrostatic and Newton's gravitational laws equivalence;
- gravitational constant dependence on electromagnetic vacuum density;
- electromagnetic nature of gravitation;
- elementary particles from electromagnetic vacuum.

Chapter 8

The Electromagnetic Field Ground State and the Cosmological Evolution*

Abstract
The vector potential quantization procedure extended to a single free-of-cavity photon leads to the zero-energy electromagnetic field ground state, the electromagnetic vacuum, overcoming the zero-point energy singularity in quantum electrodynamics. We show that the electromagnetic vacuum oscillations generate photons and that the elementary electron–positron charge is expressed precisely by the electromagnetic vacuum quantized amplitude. We also show that the gravitational constant is equally expressed through the same electromagnetic vacuum quantized amplitude, putting in evidence the electromagnetic nature of gravity and introducing the quantum vacuum influence in the cosmological equations. Fluctuations of the electromagnetic vacuum give birth to low-energy transient photons as well as to particles/antiparticles, which might be, respectively, at the origin of dark energy and dark matter in the universe.

1. Introduction

Recent astrophysical observations revealed that the universe is speeding up rather than decelerating, as it might be predicted by the

*This chapter is reproduced from C. Meis. 2018. "The Electromagnetic Field Ground State and the Cosmological Evolution", *Journal of Physics: Conference Series*, 1141, 012072.

general relativity theory in the standard matter source universe [1–6]. The cosmological constant Λ, generally identified as the quantum vacuum energy density and as the principal component of the dark energy, is advanced as the most plausible physical explanation for the cosmic acceleration [4,5,7–9]. This theoretical hypothesis leads to the well-known serious problem related to the energy scale [4, 5, 10–12]. In fact, considering a reasonable cut-off for the upper limit frequency $\sim 10^{43}$ Hz, corresponding roughly to Planck's energy of 10^{19} GeV, the theoretical value obtained for the vacuum energy density is around $10^{110}\,\mathrm{J\,m^{-3}}$. Today, we have good evidence from the astrophysical measurements that the vacuum energy density should be close to $10^{-9}\ \mathrm{J\,m^{-3}}$. This huge discrepancy between the theoretical estimation and the experimental value, of the order of 10^{120}, has been characterized as the worst prediction in the history of physics and has been given the name "vacuum catastrophe" [5, 10–12]. The various theoretical studies carried out until now [13–33] on the physics of dark energy and its relationship to the cosmological constant fail to provide a satisfactory answer to the energy scale problem.

Without using any assumptions, axioms or hypothetical models, we show here that the quantization of the vector potential of a cavity-free single photon leads naturally to the electromagnetic vacuum, a universal field with zero-energy and zero-vector potential overcoming the vacuum energy singularity. The electromagnetic vacuum is composed of an electric potential amplitude and it is expressed through the photon creation and annihilation operators involving both circular polarizations. Consequently, it is capable of generating photons (electromagnetic waves). Furthermore, it is shown that the electron–positron charge and mass are obtained exactly by the amplitude of the electromagnetic vacuum, demonstrating that electrons/positrons (leptons/antileptons) and photons are expressions of the same vacuum field. The electromagnetic vacuum fluctuations in space can generate transient states of photons and leptons/antileptons (and probably other particles/antiparticles) that might contribute to dark energy and dark matter, respectively. Considering Planck's length, we also show that the gravitational constant is equally expressed exactly through the amplitude of the electromagnetic vacuum, putting in evidence the electromagnetic character of gravity whose nature can now be studied through a new approach, probably as a Casimir effect. It is therefore deduced that the electromagnetic vacuum plays a fairly important role in the cosmological evolution.

2. The photon wave–particle equation and the photon wave function. The electromagnetic vacuum

We have shown that for a free-of-cavity k-mode photon with angular frequency ω_k, the vector potential amplitude is written as follows [13,34]:

$$\alpha_{0k} = \xi\omega_k \quad \text{with} \quad |\xi| = \left|\frac{\hbar}{4\pi e c}\right| = \frac{1}{(4\pi)^2}\left|\frac{e\mu_0}{\alpha}\right| \tag{1}$$

where ξ is a constant characterizing the intrinsic electromagnetic nature of the photon, c is the speed of light in vacuum, e is the electron–positron charge, α is the fine structure constant and μ_0 is the vacuum magnetic permeability. The vector potential of a free photon in the plane wave representation is written as

$$\begin{aligned}\vec{\alpha}_{k\lambda}(\vec{r},t) &= \omega_k\left(\xi\hat{\varepsilon}_{k\lambda}e^{i\left(\vec{k}\cdot\vec{r}-\omega_k t+\theta\right)} + \xi^*\hat{\varepsilon}^*_{k\lambda}e^{-i\left(\vec{k}\cdot\vec{r}-\omega_k t+\theta\right)}\right)\\ &= \omega_k\vec{\Xi}_{k\lambda}(\vec{r},t)\end{aligned} \tag{2}$$

where λ corresponds to the left (L) or right (R) circular polarization expressed by the complex unit vector $\hat{\varepsilon}_{k\lambda}$.

The last relation can also be written with respect to the creation and annihilation operators $a^+_{k,\lambda}$ and $a_{k,\lambda}$, respectively, for a k-mode and λ-polarization photon:

$$\begin{aligned}\tilde{\alpha}_{k\lambda}(\vec{r},t) &= \omega_k(\xi\hat{\varepsilon}_{k\lambda}a_{k\lambda}e^{i(\vec{k}\cdot\vec{r}-\omega_k t+\theta)} + \xi^*\hat{\varepsilon}^*_{k\lambda}a^+_{k\lambda}e^{-i(\vec{k}\cdot\vec{r}-\omega_k t+\theta)})\\ &= \omega_k\tilde{\Xi}_{k\lambda}(\vec{r},t)\end{aligned} \tag{3}$$

It is straightforward to demonstrate that the photon vector potential function satisfies the *vector potential–energy* (wave–particle) equation for the photon [13,34–37]:

$$i\begin{pmatrix}\xi\\ \hbar\end{pmatrix}\frac{\partial}{\partial t}\vec{\alpha}_{k\lambda}(\vec{r},t) = \begin{pmatrix}\tilde{\alpha}_{0k}\\ \tilde{H}\end{pmatrix}\vec{\alpha}_{k\lambda}(\vec{r},t) \tag{4}$$

where

$$\begin{pmatrix}\tilde{\alpha}_{0k}\\ \tilde{H}\end{pmatrix} = -i\begin{pmatrix}\xi\\ \hbar\end{pmatrix}c\vec{\nabla} \tag{5}$$

with $\tilde{H}$ the relativistic massless particle Hamiltonian with eigenvalue $\hbar\omega_k$ and $\tilde{\alpha}_{0k}$ the vector potential amplitude operator with eigenvalue $\xi\omega_k$. The insertion of the relation (5) in (4) yields the propagation equation in vacuum.

Thus, the vector potential function $\vec{\alpha}_{k\lambda}(\vec{r},t)$ with the quantized amplitude can be considered as the photon wave function in a non-local representation that can be suitably normalized.

Now, from the vector potential relations (2) and (3), we deduce that the photon wave function is mainly composed of the fundamental field $\Xi_{k\lambda}(\vec{r},t)$ times the angular frequency ω_k. Obviously, a photon subsists only for a non-zero frequency. The zero-frequency level of the electromagnetic field, i.e., the electromagnetic vacuum, corresponds to an infinite spatial entity (infinite wavelength) with zero-energy and zero-vector potential as well as zero electric and magnetic fields. It is important to remark here that this physical state lies beyond the Ehrenberg–Siday and Bohm–Aharonov situations, in which energy, electric and magnetic fields are zero, but a vector potential is still present in space [38, 39]. However, the electromagnetic vacuum does not correspond to perfect inexistence since at zero frequency, the fundamental field $\Xi_{k\lambda}(\vec{r},t)$ does not vanish but reduces to $\vec{\Xi}_{0\lambda}$, which is also written in the creation and annihilation operators representation $\tilde{\Xi}_{0\lambda}$ [13, 34, 35]:

$$\vec{\Xi}_{0\lambda} = \xi\hat{\varepsilon}_{k\lambda}e^{i\theta} + \xi^*\hat{\varepsilon}^*_{k\lambda}e^{-i\theta} \quad \tilde{\Xi}_{0\lambda} = \xi\hat{\varepsilon}_{k\lambda}a_{k\lambda}e^{i\theta} + \xi^*\hat{\varepsilon}^*_{k\lambda}a^+_{k\lambda}e^{-i\theta} \tag{6}$$

It is important to note that the electromagnetic vacuum expressed by $\Xi_{0\lambda}$ is a real field permeating all of space, has an electric potential amplitude with units Volt $\mathrm{m}^{-1}\mathrm{s}^2$ and involves both circular polarizations as well as the creation and annihilation operators. Consequently, the electromagnetic vacuum is capable of interacting with electrons (charges). Since it corresponds to a state with zero-energy and zero-vector potential, the electromagnetic vacuum complements the 'normal ordering' Hamiltonian in quantum electrodynamics. Hence, the vacuum is not an infinite sea of photons with all the wavelengths and polarizations, which leads to the well-known singularity in QED associated with the "vacuum catastrophe," but consists (probably among other components) of the electromagnetic vacuum, which is capable of generating photons of all frequencies and polarizations.

Oscillation of $\Xi_{0\lambda}$ at an angular frequency ω_k generates a photon, an integral entity extended over a wavelength with total energy

$\hbar\omega_k$ and propagating with circular polarization (R or L) with a real wave function, which is that of the vector potential $\vec{\alpha}_{k\lambda}(\vec{r},t)$. Now, the electromagnetic vacuum fluctuations imply that transient states of various photons can be generated spontaneously during a time interval. This may happen at the limits of Heisenberg's energy–time uncertainty relation $\delta E_k \cdot \delta t \sim \hbar \rightarrow \delta\alpha_{0k} \cdot \delta t \sim \xi \rightarrow \delta\omega_k \cdot \delta t \sim 1$. Obviously, the lower the frequency, the higher the lifetime δt of the transient photons generated by the electromagnetic vacuum fluctuations, and consequently, they may contribute to the long-wavelength cosmic radiation background and to its associated anisotropies.

3. The electron–positron charge and the electromagnetic vacuum

Using the fine structure constant definition $\alpha = e^2/4\pi\varepsilon_0\hbar c$, where ε_0 is the vacuum electric permittivity, we deduce directly from (1) that the electron–positron (and the proton–antiproton) elementary charge $e = \pm 1.602\ 10^{-19}$C, a fundamental physical constant, is expressed exactly through the electromagnetic vacuum amplitude ξ, the fine structure constant α and the vacuum magnetic susceptibility μ_0 [13, 40]:

$$e = (4\pi)^2 \alpha \frac{\xi}{\mu_0} \tag{7}$$

Using the relations (1) and (7) and recalling that the electron mass may be written as $m_e = e\hbar/2\mu_B$, where μ_B is the Bohr magneton, we deduce that the electron mass is also expressed exactly as a function of the electromagnetic vacuum amplitude

$$m_e = 2\pi c\, e^2 \frac{\xi}{\mu_B} \tag{8}$$

entailing that the electron mass is derived from the electromagnetic vacuum as a result of the charge and the magnetic moment.

Equations (1), (7) and (8) show that the photon vector potential as well as the electron–positron charge and mass are related directly to the electromagnetic vacuum through the amplitude ξ. This may be at the origin of the physical mechanism, governing the photon generation during the electron–positron (lepton–antilepton) annihilation and that of the electron–positron (lepton–antilepton) pair creation during the annihilation of high-energy gamma photons in the vicinity of a heavy nucleus.

We deduce that the electromagnetic vacuum fluctuations may be at the origin of the spontaneous creation of lepton–antilepton pairs (and perhaps proton–antiproton and other types of known or unknown particle–antiparticle pairs) during very short time intervals, contributing to the vacuum polarization [39].

4. The gravitational constant and the electromagnetic vacuum

It is actually well established that Planck's length ($l_P = 1.6162$ 10^{-35} m), corresponding physically to the wavelength of a photon whose energy density transforms to a black hole, appears to be a characteristic distance in the space–time continuum [4, 9, 12]. At the present state of knowledge and technology, Planck's length cannot be measured experimentally, and it is generally accepted that below this characteristic distance, physical notions have no sense. It is easily established that the gravitational constant $G = 6.674\ 10^{-11}$ $\mathrm{m^3\,kg^{-1}s^{-2}}$ can be expressed exactly using Planck's length l_P and the electromagnetic vacuum amplitude ξ:

$$G = \frac{1}{4\pi\epsilon_0\alpha}\left(\frac{l_P}{4\pi\xi}\right)^2 \tag{9}$$

Consequently, the electromagnetic character of gravity lies in the nature of the electromagnetic vacuum, opening perspectives for the study of the gravitational field through a new approach. In fact, following the relations (2) and (3), any charge undergoing a periodic motion at a given angular frequency ω experiences in its frame the presence of "vacuum photons" due to the electromagnetic vacuum. For a physical system characterized by a mean electron density $< n_e >$, it is usual to attribute a plasma frequency ω_p, which is written as

$$\omega_p = e < n_e >^{1/2} (\varepsilon_0 m_e)^{-1/2} \tag{10}$$

Therein, it might be interesting to investigate whether gravity is an analogue of the Casimir effect due to the electromagnetic vacuum radiation pressure at roughly the plasma frequencies felt by the charge densities of the bodies in their own frame.

5. The electromagnetic vacuum and the Cosmological Equations

The field equations of general relativity applied to the Friedmann–Lemaitre–Robertson–Walker (FLRW) space–time metric

$$ds^2 = c^2dt^2 - a^2(t)\left(\frac{dr^2}{1-kr^2} + r^2d\theta^2 + r^2\sin^2\theta\, d\phi^2\right) \tag{11}$$

lead to the FLRW equations, presently considered as the principal equations of cosmology [4, 5]:

$$\frac{1}{c^2}\left(\frac{1}{a(t)}\frac{d\,a(t)}{dt}\right)^2 + \frac{k}{R_0^2a^2(t)} = \frac{8\pi\, G}{3\, c^4}\rho + \frac{\Lambda}{3} \tag{12}$$

$$\frac{1}{c^2}\frac{1}{a(t)}\frac{d^2a(t)}{dt^2} = -\frac{4\pi\, G}{3\, c^4}(\rho + 3P) + \frac{\Lambda}{3} \tag{13}$$

where the factor $a(t)$ characterizes the cosmic expansion ($a(t) = 1$ in our epoch), k denotes the three-dimensional space curvature (-1 for negative curvature, 0 for flat Euclidean space, 1 for positive curvature), $\{r, \theta, \phi\}$ are the coordinates in the spherical system, ρ is the total energy density of the universe composed of all the components (matter, radiation, dark matter, dark energy, etc.), P is the total pressure of the components, R_0 is the present radius of the space curvature and Λ is the cosmological constant.

The last one was introduced by Einstein in the general relativity equations in order to obtain a static model of the universe. Shortly after, the astronomical observations confirmed the expansion of the universe, and the cosmological constant was abandoned. In recent years, the cosmic expansion acceleration has again justified its reintroduction. Today, we have good evidence that the cosmological constant should correspond to the vacuum energy density ρ_{vacuum}, resulting in a negative vacuum pressure P_{vacuum} expressed as follows [5]:

$$\Lambda\, c^4 = 8\pi\, G\, \rho_{\text{vacuum}} = -8\pi\, G\, P_{\text{vacuum}} \tag{14}$$

Denoting by ρ_{Ξ_0} the energy density due to the electromagnetic vacuum fluctuations generating transient photons, we may assume that

$\rho_{\text{vacuum}} \approx \rho_{\Xi_0}$, so that taking into account the relations (9) and (14), the FLRW equations are written in terms of the electromagnetic vacuum amplitude ξ, the electron charge e and Planck's length l_P:

$$\left(\frac{1}{a(t)}\frac{da(t)}{dt}\right)^2 + \frac{k\,c^2}{R_0^2\,a^2(t)} = \frac{2}{3}\frac{l_P^2}{|e\xi|}\left[\rho + \rho_{\Xi_0}\right] \tag{15}$$

$$\frac{1}{a(t)}\frac{d^2a(t)}{dt^2} = -\frac{2}{3}\frac{l_P^2}{|e\,\xi|}\left[\frac{1}{2}(\rho + 3P) - \rho_{\Xi_0}\right] \tag{16}$$

where ρ and P also involve the contributions of the electromagnetic vacuum fluctuations contributing to dark energy and dark matter. Apparently, the influence of the electromagnetic vacuum upon the cosmic evolution appears to be fairly important, linking quantum electrodynamics to cosmology.

6. Conclusion

The vector potential amplitude quantization for a single free-of-cavity photon leads to the definition of a zero-energy and zero-vector potential field $\Xi_{0\lambda}$ permeating all of space and identified as a vacuum component with the real amplitude $\xi = e\mu_0/(4\pi)^2\alpha$. Photons, and thus electromagnetic waves, are oscillations of this vacuum field. Fluctuations of the electromagnetic vacuum generate low-energy transient photons that may contribute to the observed vacuum energy density considered responsible for the cosmic acceleration.

The lepton–antilepton charge is expressed exactly through the electromagnetic vacuum amplitude ξ demonstrating the physical relationship between photons and leptons/antileptons. It is worthwhile to investigate further this aspect in order to understand the mechanisms governing the photon transformation to leptons/antileptons and inversely. Transient pairs of leptons/antileptons and probably other pairs of known (or unknown) particles/antiparticles may be issued from the electromagnetic vacuum fluctuations at the limits of Heisenberg's energy-time uncertainty that might contribute to the vacuum polarization in the universe.

We have finally shown that the gravitational constant is equally expressed through the electromagnetic vacuum amplitude, entailing that the electromagnetic character of gravity can now be studied through a new approach.

To conclude, according to all the above considerations, we have expressed the contribution of the electromagnetic vacuum in the FLRW cosmological equations, revealing the decisive role that it plays in the cosmic evolution.

References

[1] A. G. Riess *et al.*, Observational evidence from supernovae for an accelerating universe and a cosmological constant, *Astron. J.* 116, 1009 (1998).

[2] S. Perlmutter *et al.*, Measurements of Ω and Λ from 42 High Redshift Supernovae, *Astrophys. J.* 517, 565 (1999).

[3] R. F. L. Holanda and S. H. Pereira, Can galaxy clusters, type Ia supernovae, and the cosmic microwave background rule out a class of modified gravity theories, *Phys. Rev. D* 94, 104037 (2016).

[4] V. Mukhanov, *Physical Foundations of Cosmology* (Cambridge University Press, Cambridge, UK, 2005).

[5] J. A. Frieman, M. S. Turner, and D. Huterer, Dark energy and the accelerating Universe, *Annu. Rev. Astron. Astrophys.* 46, 385 (2008).

[6] E. V. Linder, Mapping the cosmological expansion, *Rep. Prog. Phys.* 71, 056901 (2008).

[7] C. Beck, Axiomatic approach to the cosmological constant, *Physica A.* 388, 3384 (2009).

[8] P. J. E. Peebles and B. Ratra, The cosmological constant and dark energy, *Rev. Mod. Phys.* 75, 559 (2003).

[9] T. Hey and P. Walters, *The New Quantum Universe* (Cambridge University Press, Cambridge, UK, 2003).

[10] R. J. Adler, B. Casey, and O. C. Jacob, Vacuum catastrophe: An elementary exposition of the cosmological constant problem, *Am. J. Phys.* 63, 620 (1995).

[11] T. Padmanabhan, Cosmological constant: the weight of the vacuum, *Phys. Rep.* 380, 235 (2003).

[12] M. P. Hobson, G. P. Efstathiou, and A. N. Lasenby, *General Relativity: An Introduction for Physicists* (Cambridge University Press, Cambridge, UK, 2006).

[13] C. Meis, *Light and Vacuum*, 2nd ed. (World Scientific, Singapore, 2017).

[14] A. Sangwan, A. Mukherjee, and H. K. Jassal, Reconstructing the dark energy potential, *J. Cosmol. Astropart. Phys.* 01, 018 (2018).
[15] R. Caldwell and E. V. Linder, The Limits of quintessence, *Phys. Rev. Lett.* 95, 141301 (2005).
[16] I. Zlatev, L. Wang, and P. J. Steinhardt, Quintessence, cosmic coincidence, and the cosmological constant, *Phys. Rev. Lett.* 82, 896 (1999).
[17] C. Armendariz-Picon, V. F. Mukhanov, and P. J. Steinhardt, A dynamical solution to the problem of a small cosmological constant and late-time cosmic acceleration, *Phys. Rev. Lett.* 85, 4438 (2000).
[18] J. Garriga and V. F. Mukhanov, Perturbations in k-inflation, *Phys. Lett. B* 458, 219 (1999).
[19] E. Babichev, V. Mukhanov, and A. Vikman, k-Essence, superluminal propagation, causality and emergent geometry, *J. High Energy Phys.* 02, 101 (2008).
[20] C. Wetterich, The cosmon model for an asymptotically vanishing time-dependent cosmological constant, *Astron. Astrophys.* 301, 321 (1995).
[21] L. Amendola, Coupled quintessence, *Phys. Rev. D.* 62, 043511 (2000).
[22] R.-G. Cai and A. Wang, Cosmology with interaction between phantom dark energy and dark matter and the coincidence problem, *J. Cosmol. Astropart. Phys.* 03, 002 (2005).
[23] E. J. Copeland, A. R. Liddle, and D. Wands, Exponential potentials and cosmological scaling solutions, *Phys. Rev. D* 57, 4686 (1998).
[24] A. Y. Kamenshchik, U. Moschella, and V. Pasquier, An alternative to quintessence, *Phys. Lett. B.* 511, 265 (2001).
[25] S. Capozziello, S. Carloni, and A. Troisi, Quintessence without scalar fields, *Recent Res. Dev. Astron. Astrophys.* 1, 625 (2003).
[26] M. C. Bento, O. Bertolami, and A. A. Sen, Generalized Chaplygin gas, accelerated expansion and dark energy-matter unification, *Phys. Rev. D.* 66, 043507 (2002).
[27] S. Nojiri and S. D. Odintsov, Introduction to modified gravity and gravitational alternative for dark energy, *Int. J. Geom. Methods Mod. Phys.* 4, 115 (2007).
[28] Y.-S. Song, W. Hu, and I. Sawicki, The large-scale structure of f(R) gravity, *Phys. Rev. D.* 75, 044004 (2007).
[29] S. Tsujikawa, Matter density perturbations and effective gravitational constant in modified gravity models of dark energy, *Phys. Rev. D.* 76, 023514 (2007).
[30] S. Tsujikawa, Observational signatures of f(R) dark energy models that satisfy cosmological and local gravity constraints, *Phys. Rev. D.* 77, 023507 (2008).

[31] H. Alnes, M. Amarzguioui, and Ø. Grøn, An inhomogeneous alternative to dark energy? *Phys. Rev. D.* 73, 083519 (2006).
[32] H. Iguchi, T. Nakamura, and K.-I. Nakao, Is dark energy the only solution to the apparent acceleration of the present universe? *Prog. Theor. Phys.* 108, 809 (2002).
[33] A. A. Starobinsky, A new type of isotropic cosmological models without singularity, *Phys. Lett. B.* 91, 99 (1980).
[34] C. Meis, Vector potential quantization and the quantum vacuum, *Phys. Res. Int.* 2014, 187432 (2014).
[35] C. Meis and P. R. Dahoo, Vector potential quantization and the photon wave-particle representation, *J. Phys.: Conf. Ser.* 738, 012099 (2016). DOI: 10.1088/1742-6596/738/1/012099.
[36] C. Meis and P. R. Dahoo, Vector potential quantization and the photon intrinsic electromagnetic properties: Towards nondestructive photon detection, *Int. J. Quantum Inf.* 15, 1740003 (2017). DOI: 10.1142/S0219749917400032.
[37] C. Meis and P. R. Dahoo, Quantized vector potential and the photon wave function, *J. Phys.: Conf. Ser.* 936, 012004 (2017). DOI: 10.1088/1742-6596/936/1/012004.
[38] W. Ehrenberg and R. E. Siday, The refractive index in electron optics, *Proc. Phys. Soc. London, Sect. B.* 62, 8 (1949).
[39] Y. Aharonov and D. Bohm, Significance of electromagnetic potentials in the quantum theory, *Phys. Rev.* 115, 485 (1959).
[40] C. Meis and P. R. Dahoo, *AIP Conf. Proc.* 2040, 020011 (2018).

Chapter 9

The Vacuum Electromagnetic Energy Density: Toward a Physical Comprehension of the Vacuum Energy Scale Problem*

Abstract

The electromagnetic vacuum, a zero-energy cosmic field permeating all of space, issues naturally from Maxwell's electromagnetic theory, and its fluctuations confer a physical solution to the vacuum energy scale problem. They are more significant at long wavelengths, underlying principally the low frequency up to infrared cosmic radiation background. Assuming an exponentially decreasing probability with increasing frequency for the fluctuating photons, the obtained vacuum energy density is in agreement with the astrophysical observations. An equivalent treatment can be applied to all other fields. Consequently, the electromagnetic vacuum fluctuations might be the principal component of the dark energy considered responsible for the observed cosmic acceleration.

*This chapter is reproduced from C. Meis. 2020. "The Vacuum Electromagnetic Energy Density: Towards a Physical Comprehension of The Vacuum Energy Scale Problem", *International Journal of Modern Physics A*, 35(25), 2050153.

1. Introduction

The interpretation of the observed cosmic expansion acceleration [1,2] based on dark energy, generally identified as the vacuum energy issued from the quantum field theory, has generated major theoretical difficulties in physics related to the energy scale widely known as "vacuum catastrophe" [3–11]. Dark energy is the main component of the universe, although its nature is completely unknown.

A large number of independent investigations employing various methods during the last two decades have confirmed the cosmic acceleration [12–32]. The universe is speeding up rather than decelerating, contradicting the predictions of general relativity based on the standard matter source universe. Consequently, a low energy density field should fill up the whole space, pushing the galaxies to move away. The cosmological constant Λ identified as the quantum vacuum energy density and the main component of dark energy seems to be the most plausible physical explanation [7,8,11]. In fact, many studies based on quantum chromodynamics (QCD) and quantum electrodynamics (QED) have been carried out in order to investigate the contribution of the quantum vacuum to the cosmological constant. Our knowledge of the QCD vacuum is relatively elusive; conversely, QED offers a more comprehensive framework for understanding the quantum vacuum energy density scale problem [3, 5, 7] that we analyze briefly before advancing to an alternative solution.

In QED formalism, the electromagnetic energy density of the vacuum in a volume V is calculated by the well-known expression [33]

$$\rho_{\text{vacuum}} = \frac{1}{V} \sum_{k,\lambda} \frac{1}{2} \hbar \omega_k \tag{1}$$

where the summation runs over all the angular frequencies ω_k of the k-mode photons, λ takes only two values corresponding to circular left and right polarizations and $\hbar$ is Planck's reduced constant. The discrete summation is replaced by a continuous one by employing the following transformation relation issued from the density of states theory:

$$\sum_{k,\lambda} \rightarrow \frac{V}{\pi^2 c^3} \int \omega^2 d\omega \tag{2}$$

where c is the velocity of light in vacuum.

Thus, the electromagnetic vacuum energy density is written as

$$\rho_{\text{vacuum}} = \frac{\hbar}{2\pi^2 c^3} \int_0^\infty \omega^3 d\omega \tag{3}$$

which is infinite at any point in space [33].

From a physical point of view and in order to avoid the singularity, one may reasonably choose the upper limit of the integration to be fixed at the value $\omega_{\text{max}}/2\pi \sim 10^{43}$ Hz, which is the angular frequency corresponding to Planck's energy $E_{\text{Planck}} \sim 10^{19}$ GeV. In fact, at the present stage of knowledge, Planck's energy is considered to be the highest possible for a single photon. Under this condition, the value obtained for the electromagnetic vacuum energy density in QED is roughly $\rho_{\text{vacuum}} \sim 10^{110}\,\text{J}\,\text{m}^{-3}$. When taking into account all the other fields (leptons, mesons, etc.), this value gets higher, though the order of magnitude does not really change significantly [7].

On the other hand, recent well-validated astrophysical observations have shown that the vacuum energy density should be close to $\rho_{\text{vacuum}}^{\text{observed}} \sim 10^{-9}\,\text{J}\,\text{m}^{-3}$, which is roughly 120 orders of magnitude lower than the estimated one in QED [5–10]. This "astronomic" discrepancy between the theoretical and observed values is the worst ever met in the history of physics and constitutes the vacuum energy scale problem, also called the vacuum catastrophe [3–7].

2. The electromagnetic field ground state and the vacuum energy density

A dimension analysis of the general solution of the vector potential, obtained from Maxwell's equations, implies that its amplitude is proportional to the angular frequency. Consequently, for a cavity-free k-mode photon, we can write the vector potential amplitude as follows [34–38]:

$$\alpha_{0k}(\omega_k) = \xi\omega_k \tag{4}$$

It is important to note that the last relation is neither a hypothesis nor a suggested model but issues naturally from Maxwell's electromagnetic theory.

The vector potential amplitude quantization constant ξ has been evaluated by normalizing the energy of a single k-mode plane electromagnetic wave over a wavelength λ_k to the photon energy $E_k = \hbar\omega_k$, obtaining

$$\xi = \frac{\hbar}{4\pi ec} = \frac{e\mu_0}{\alpha(4\pi)^2} \simeq \pm 1.747\,10^{-25}\,\mathrm{V\,m^{-1}\,s^2} \tag{5}$$

where $\alpha \simeq 1/137$ is the fine structure constant, e is the electron–positron charge and μ_0 is the vacuum magnetic permeability.

Hence, for the k-mode photon, the vector potential function $\vec{\alpha}_k(\vec{r},t)$ is

$$\vec{\alpha}_{k\lambda}(\vec{r},t) = \xi\omega_k[\hat{\varepsilon}_{k\lambda}e^{i(\vec{k}\cdot\vec{r}-\omega_k t+\theta)} + cc] = \omega_k\vec{\Xi}_{k\lambda}(\vec{r},t) \tag{6}$$

where $\hat{\varepsilon}_{k\lambda}$ is the circular polarization complex unit vector, $\vec{k}$ is the wave vector with amplitude $|\vec{k}| = 2\pi/\lambda_k$, λ_k being here the wavelength of the mode k, θ is a phase parameter and cc is the complex conjugate.

The function $\vec{\alpha}_{k\lambda}(\vec{r},t)$ satisfies the *vector potential–energy (wave–particle)* equation for the photon [36,37] composed of the Schrödinger equation for the energy with the relativistic massless particle Hamiltonian $\tilde{H} = -i\hbar c\vec{\nabla}$ coupled to a symmetrical relation for the vector potential with the vector potential amplitude operator $\tilde{\alpha}_{0k} = -i\xi c\vec{\nabla}$:

$$i\begin{pmatrix}\xi\\ \hbar\end{pmatrix}\frac{\partial}{\partial t}\vec{\alpha}_{k\lambda}(\vec{r},t) = \begin{pmatrix}\tilde{\alpha}_{0k}\\ \tilde{H}\end{pmatrix}\vec{\alpha}_{k\lambda}(\vec{r},t) \tag{7}$$

Thus, in agreement with the experimental evidence, a k-mode photon is a wave–corpuscle entity expressed as a quantum of the electromagnetic field extended over a wavelength, with energy $E_k = \hbar\omega_k$, momentum $|\vec{p}_k| = E_k/c$, with circular left or right polarization and corresponding spin $\pm\hbar$, guided by a non-local vector potential wave function.

From the vector potential, it is directly drawn that the electric field amplitude of a cavity-free k-mode photon is proportional to the square of the angular frequency [36]

$$|\vec{\varepsilon}_{k\lambda}(\vec{r},t)| = |\partial\vec{\alpha}_{k\lambda}(\vec{r},t)/\partial t| \propto \xi\omega_k^2 \tag{8}$$

Now, when $\omega \to 0$, both the photon energy and the vector potential tend to zero. However, for $\omega = 0$, the ground state does not correspond to a perfect empty space because the function $\vec{\Xi}_{k\lambda}(\vec{r},t)$ does not vanish and tends to the field $\Xi_{0\lambda}$, which is written as [34, 38–40]

$$\vec{\Xi}_{0\lambda} = \xi[\hat{\varepsilon}_{k\lambda}e^{i\theta} + \hat{\varepsilon}^*_{k\lambda}e^{-i\theta}] \tag{9}$$

Obviously, $\vec{\Xi}_{0\lambda}$ is a real field extending to infinity ($\lambda_k \to \infty$). In total absence of energy, it is a component of the vacuum having an electric potential dimension. Thus, the electromagnetic field ground state is an underlying background field permeating all of space, i.e., the electromagnetic vacuum.

This formalism confers a basic property to the quantum vacuum and complements the QED concept when considering the normal ordering Hamiltonian, overcoming the infinite vacuum energy singularity. It leads to a coherent link between the classical electromagnetic wave theory issued from Maxwell's equations and QED [36, 37, 39].

Heisenberg's energy–time uncertainty relation applied in Eq. (4) entails directly that the vector potential amplitude is also subject to a fluctuation uncertainty:

$$\delta E_k \cdot \delta t \approx \hbar \implies \delta\alpha_{0k} \cdot \delta t \approx \xi \implies \delta\omega_k \cdot \delta t \approx 1 \tag{10}$$

Fluctuations of the electromagnetic vacuum in space may give birth to transient states of photons underlying the cosmic radiation background [40].

From the relation (10), it becomes evident that the probability P_k for the spontaneous generation of a k-mode transient photon by the electromagnetic vacuum fluctuations decreases with increasing frequency. Hence, we may reasonably assume that P_k depends on the electric field amplitude $|\vec{\varepsilon}_k|$ of the k-mode photon generated in vacuum, following the expression

$$P_k = p_0 e^{-|\vec{\varepsilon}_k|/|\vec{\varepsilon}_m|} = p_0 e^{-\omega_k^2/\omega_m^2} \tag{11}$$

where $|\vec{\varepsilon}_m|$ corresponds to a mean value of the photon electric field in the electromagnetic spectrum. From the relation (10), the spontaneous generation probability for photons at extremely low frequencies should be close to one, so that we have $p_0 \sim 1$.

Weighting the number of states by the probability P_k per state, the discrete summation transformation in Eq. (2) becomes

$$\sum_{k,\lambda} \to \frac{V}{\pi^2 c^3} \int p_0 e^{-\omega^2/\omega_m^2} \omega^2 d\omega \tag{12}$$

Hence, using the relation (12) instead of (2), the electromagnetic vacuum energy density now becomes finite and is written as

$$\rho_{\text{vacuum}} \sim \frac{\hbar}{\pi^2 c^3} \int_0^{\infty} p_0 e^{-\omega^2/\omega_m^2} \omega^3 d\omega = p_0 \frac{\hbar \omega_m^4}{2\pi^2 c^3} \tag{13}$$

According to the uncertainty relation (10), the very energetic gamma photons corresponding to frequencies higher than $\sim 10^{25}$ Hz have an extremely low probability to be spontaneously generated by the electromagnetic vacuum fluctuations. Consequently, the mean angular frequency ω_m of the electromagnetic fluctuations could be roughly in the logarithmic middle of the electromagnetic spectrum, so that $\omega_m/2\pi \propto 10^{12}\,\text{Hz}$ ($\lambda_\kappa = 300\,\mu\text{m}$). Using the last value for ω_m in Eq. (13), we obtain a vacuum electromagnetic energy density $\rho_{\text{vacuum}} \sim 3 \cdot 10^{-10}\,\text{J}\,\text{m}^{-3}$.

Taking into account that we have considered only the electromagnetic field, this result is in good agreement with the astrophysical observations. We may consider an analogue statement for all the other k-mode fields [7], which also emerge from the electromagnetic vacuum [40], weighting the integration by e^{-k^2/k_m^2}, so that the total vacuum energy density becomes finite.

The photons issued from the electromagnetic vacuum fluctuations underlie the cosmic radiation background, decreasing strongly at high frequencies. In fact, the relation (10) entails that the lifetime should be longer for the lower frequency photons, implying a significant contribution to the cosmic long-wavelength radio background. Indeed, this is compatible with ARCADE-2 experimental results that have shown an unexpected increasing intensity at lower frequencies in the interval studied from 10.5 GHz to 0.022 GHz [41, 42]. It could be of very high interest to investigate the cosmic radiation background at even lower frequencies.

The energy density in the interval $d\omega$in Eq. (13) is expressed by the function $\omega^3 e^{-\omega^2/\omega_m^2}$, which has a maximum value for $\omega = \sqrt{3/2}\omega_m$. Consequently, we may expect that significant energy density fluctuations could underlie the cosmic infrared background (CIB). Rocket expedition measurements for the investigation of the spatial infrared spectrum have first reported evidence of a CIB [43–45]. Some of the measurements have entailed that the integrated energy density of the CIB might exceed that of the cosmic microwave background (CMB) [46,47]. Indeed, IRAS observations of the diffuse infrared background in the range of 10–30 μm [48, 49] have shown intensities comparable to those of the CMB. Conversely, investigations carried out with SPITZER [50] have permitted fixing low limits for the CIB which are much lower than the IRAS values, entailing very high uncertainties in this range of frequencies. This might be due to the significant electromagnetic vacuum fluctuations in the IR range shown above. Despite the challenging difficulties for discriminating and removing the strong infrared signals, coming from the solar system and galactic sources, it is of crucial importance to carry out precise measurements in the range of 1–500 μm in order to have a clear idea of the influence of the electromagnetic field ground state fluctuations upon the CIB.

3. Conclusion

Fluctuations of the electromagnetic field ground state $\vec{\Xi}_{0\lambda}$, i.e., the electromagnetic vacuum, may generate transient photons underlying the cosmic radiation background. Assuming an exponential decrease of the generation probability with increasing photon electric field amplitude, the electromagnetic vacuum energy density summation becomes finite and comparable to the observed values, conferring a natural solution to the vacuum energy scale problem. Consequently, the electromagnetic vacuum fluctuations may be a component of dark energy. The lifetime of the transient photons is longer at low frequencies; consequently, a detailed experimental exploration of the very low radiofrequency cosmic background could provide useful information. The tail of the electromagnetic vacuum fluctuations toward higher frequencies underlies CMB, CIB as well as COB. The more significant fluctuations should underlie the cosmic THz frequencies and CIB, in which indeed higher uncertainties have already been observed.

References

[1] A. G. Riess *et al.*, Observational evidence from supernovae for an accelerating universe and a cosmological constant, *Astron. J.* 116, 1009–1038 (1998).

[2] S. Perlmutter *et al.*, Measurements of omega and lambda from 42 high-redshift supernovae, *Astrophys. J.* 517, 565–586 (1999).

[3] R. J. Adler, B. Casey, and O. C. Jacob, Vacuum catastrophe: An elementary exposition of the cosmological constant problem, *Am. J. Phys.* 63, 620–626 (1995).

[4] S. Weinberg, The cosmological constant problem, *Rev. Mod. Phys.* 61, 1 (1989).

[5] T. Padmanabhan, Cosmological constant: The weight of the vacuum, *Phys. Rep.* 380, 235 (2003).

[6] M. P. Hobson, G. P. Efstathiou, and A. N. Lasenby, *General Relativity: An Introduction for Physicists* (Cambridge University Press, Cambridge, UK, 2006).

[7] J. A. Frieman, M. S. Turner, and D. Huterer, Dark energy and the accelerating universe, *Annu. Rev. Astron. Astrophys.* 46, 385–432 (2008).

[8] L. Labun and J. Rafelski, Vacuum structure and dark energy, *Int. J. Mod. Phys. D.* 19, 2299 (2010).

[9] A. Silvestri and M. Trodden, Approaches to understanding cosmic acceleration, *Rep. Prog. Phys.* 72, 096901 (2009).

[10] E. V. Linder, Mapping the cosmological expansion, *Rep. Prog. Phys.* 71, 056901 (2008).

[11] E. J. Copeland, M. Sami, and S. Tsujikawa, Dynamics of dark energy, *Int. J. Mod. Phys. D.* 15, 1753–1936 (2006).

[12] A. G. Riess *et al.*, Type Ia supernova discoveries at z >1 from the Hubble Space Telescope: Evidence for past deceleration and constraints on dark energy evolution, *Astrophys. J.* 607, 665–687 (2004).

[13] D. Scolnic *et al.*, SUPERCAL: Cross-calibration of multiple photometric systems to improve cosmological measurements with type Ia supernovae, *Astrophys. J.* 815(2), 117 (2015).

[14] R. A. Knop *et al.*, New constraints on Ω_M, Ω_Λ, and w from an independent set of eleven high-redshift supernovae observed with HST, *Astrophys. J.* 598, 102 (2003).

[15] A. G. Riess *et al.*, New Hubble Space Telescope discoveries of Type Ia supernovae at z >1: Narrowing constraints on the early behavior of dark energy, *Astrophys. J.* 659, 98–121 (2007).

[16] D. N. Spergel *et al.*, First-year Wilkinson Microwave Anisotropy Probe (WMAP) observations: Determination of cosmological parameters, *Astrophys. J. Suppl. Ser.* 148, 175 (2003).

[17] D. N. Spergel *et al.*, Wilkinson Microwave Anisotropy Probe (WMAP) three-year results: Implications for cosmology, *Astrophys. J. Suppl. Ser.* 170, 377 (2007).

[18] E. Komatsu *et al.*, Seven-year Wilkinson Microwave Anisotropy Probe (WMAP) observations: Cosmological interpretation, *Astrophys. J. Suppl. Ser.* 192, 18 (2011).
[19] P. de Bernardis *et al.*, A flat universe from high-resolution maps of the cosmic microwave background radiation, *Nature* 404, 955–959 (2000).
[20] P. S. Corasaniti, M. Kunz, D. Parkinson, E. J. Copeland, and B. A. Bassett, The foundations of observing dark energy dynamics with the Wilkinson Microwave Anisotropy Probe, *Phys. Rev. D.* 70, 083006 (2004).
[21] R. G. Crittenden, P. Natarajan, U.-L. Pen, and T. Theuns, Discriminating weak lensing from intrinsic spin correlations using the curl-gradient decomposition, *Astrophys. J.* 568, 20 (2002).
[22] S. Asaba, C. Hikage, K. Koyama, K. Zhao, and A. Hojjati, Principal component analysis of modified gravity using weak lensing and peculiar velocity measurements, *J. Cosmol. Astropart. Phys.* 08, 029 (2013).
[23] P. Zhang, M. Liguori, R. Bean, and S. Dodelson, Probing gravity at cosmological scales by measurements which test the relationship between gravitational lensing and matter overdensity, *Phys. Rev. Lett.* 99, 141302 (2007).
[24] R. A. Vanderveld, M. J. Mortonson, W. Hu, and T. Eifler, Testing dark energy paradigms with weak gravitational lensing, *Phys. Rev. D.* 85, 103518 (2012).
[25] D. J. Eisenstein *et al.*, Detection of the baryon acoustic peak in the large-scale correlation function of SDSS luminous red galaxies, *Astrophys. J.* 633, 560–574 (2005).
[26] W. J. Percival *et al.*, Measuring the baryon acoustic oscillation scale using the SDSS and 2dFGRS, *Mon. Not. R. Astron. Soc.* 381, 1053–1066 (2007).
[27] C. Blake *et al.*, The WiggleZ Dark Energy Survey: Mapping the distance-redshift relation with baryon acoustic oscillations, *Mon. Not. R. Astron. Soc.* 418, 1707–1724 (2011).
[28] L. Anderson *et al.*, The clustering of galaxies in the SDSS-III Baryon Oscillation Spectroscopic Survey: Baryon acoustic oscillations in the Data Releases 10 and 11 Galaxy samples, *Mon. Not. R. Astron. Soc.* 441(1), 24–62 (2014).
[29] S. W. Allen, A. E. Evrard, and A. B. Mantz, Cosmological parameters from observations of galaxy clusters, *Annu. Rev. Astron. Astrophys.* 49, 409–470 (2011).
[30] S. Basilakos and L. Perivolaropoulos, Testing GRBs as standard candles, *Mon. Not. R. Astron. Soc.* 391, 411–419 (2008).

[31] A. Mantz, S. W. Allen, D. Rapetti, and H. Ebeling, The observed growth of massive galaxy clusters I: Statistical methods and cosmological constraints, *Mon. Not. R. Astron. Soc.* 406, 1759–1772 (2010).
[32] Y. Wang, Model-independent distance measurements from gamma-ray bursts and constraints on dark energy, *Phys. Rev. D* 78, 123532 (2008).
[33] P. W. Milonni, *The Quantum Vacuum* (Academic Press, London, UK, 1994).
[34] C. Meis, Vector potential quantization and the quantum vacuum, *Phys. Res. Int.* 2014, 187432 (2014).
[35] C. Meis, *Light and Vacuum*, 2nd ed. (World Scientific, Singapore, 2017).
[36] C. Meis and P. R. Dahoo, Vector potential quantization and the photon intrinsic electromagnetic properties: Towards nondestructive photon detection, *Int. J. Quantum Inf.* 15, 1740003 (2017).
[37] C. Meis and P. R. Dahoo, Quantized vector potential and the photon wave function, *J. Phys.: Conf. Ser.* 936, 012004 (2017).
[38] C. Meis and P. R. Dahoo, The single photon state, the quantum vacuum and the elementary electron/positron charge, *AIP Conf. Proc.* 2040, 020011 (2018).
[39] C. Meis, Quantized field of single photons, in *Single Photon Manipulation*, K. Xia, Ed. (IntechOpen, London, UK, 2019), pp. 15–36.
[40] C. Meis, The electromagnetic field ground state and the cosmological evolution, *J. Phys.: Conf. Ser.* 1141, 012072 (2018).
[41] T. Vernstrom, S. Douglas, J. V. Wall, Contribution to the diffuse radio background from extragalactic radio sources, *Mon. Not. R. Astron. Soc.* 415, 3641–3649 (2011).
[42] A. Kogut *et al.*, ARCADE 2 observations of galactic radio emission, *Astrophys. J.* 734, 4 (2011).
[43] M. Noda *et al.*, Rocket observation of the near-infrared spectrum of the sky, *Astrophys. J.* 391, 456 (1992).
[44] M. Kawada *et al.*, A rocket-borne observation of the far-infrared sky at high galactic latitude, *Astrophys. J.* 425, 89 (1994).
[45] J. L. Puget *et al.*, Tentative detection of a cosmic far-infrared background with COBE, *Astron. Astrophys.* 308, L5 (1996).
[46] M. T. Ressell and M. S. Turner, *Comments Astrophys.* 14, 323 (1990).
[47] M. G. Hauser, Unveiling the CIB, in *The Cosmic Infrared Background*, A. M. Symonds, Ed., AIP Conf. Proc. No. 348 (AIP, New York, NY, USA, 1996), p. 11.
[48] M. G. Hauser *et al.*, IRAS observations of the diffuse infrared background, *Astrophys. J.* 278, L15–L18 (1984).

[49] M. G. Hauser *et al.*, The COBE diffuse infrared background experiment search for the cosmic infrared background. I. Limits and detections, *Astrophys. J.* 508, 25–43 (1998).
[50] H. Dole *et al.*, The cosmic infrared background resolved by Spitzer, *Astron. Astrophys.* 451, 417–429 (2006).

Chapter 10

Photocosmos: Is the Universe Made of Light? Photons, Particles and Gravitation form the Quantum Vacuum*

Abstract
Without stating postulates or making any hypothesis, the quantization of the vector potential amplitude at a single photon level naturally yields the electromagnetic field ground state, corresponding to the electromagnetic vacuum having both classical and quantum representations. It is a zero-energy cosmic field with an electric nature permeating all of space and composed of real quantum states, called here *kenons* (from $\kappa\varepsilon\nu o$ = vacuum). It overcomes the vacuum energy singularity in quantum electrodynamics without compromising any of the achievements. Photons appear clearly as local oscillations of the electromagnetic vacuum states, propagating at the speed of light and having a non-local real wave function. The elementary positive and negative charges derive equally from the electromagnetic vacuum and may be assimilated to standing photon states in specific topological configurations. Furthermore, it is shown that the masses of all elementary particles and antiparticles derive from the electromagnetic vacuum and are related to the electron–positron elementary charge. The mass effect results from the charge states and their magnetic moments. All neutral particles are composed of positive and negative charges. The electromagnetic nature of

*This chapter is reproduced from C. Meis. 2025. "Photocosmos: Is the Universe Made of Light? Photons, Particles and Gravitation form the Quantum Vacuum", *Journal of High Energy Physics, Gravitation and Cosmology*, 11(2), pp. 209–223.

particles implies that Newton's and Coulomb's laws are naturally equivalent. The gravitational constant also has an electromagnetic nature and depends on the electromagnetic vacuum density of states, entailing that it may not be universal. The electromagnetic vacuum is the source of light, matter and gravitational effects in the universe.

1. Introduction

Following the publication of Maxwell's equations in 1865, the vector potential $\vec{A}(\vec{r},t)$ was considered to be simply a mathematical function, with no physical existence [1,2]. It permits us to calculate the electric $\vec{E}(\vec{r},t) = -\partial\vec{A}(\vec{r},t)/\partial t$ and magnetic $\vec{B}(\vec{r},t) = \vec{\nabla} \times \vec{A}(\vec{r},t)$ fields of the electromagnetic waves, generally considering the scalar potential to be constant in space [3]. In 1949, Ehrenberg and Siday put in evidence experimentally the direct influence of the vector potential on charges [4]. This was also confirmed later by Chambers (1960), followed by the works of Tonomura (1982) and Osakabe (1986) [5–7]. Hence, the physical reality of the vector potential and its direct influence on charges in the absence of electric and magnetic fields has been demonstrated experimentally. It is well established that the electric and magnetic fields derive from the vector potential, which is a real field and constitutes the essence of the electromagnetic waves [3,8–10].

Thus, the quantization of the electromagnetic field in QED is based precisely on the vector potential [11–13]. In recent years, the vector potential quantization has been enhanced to a single photon level, putting in evidence the electromagnetic vacuum, a real field having an equivalent representation in classical and quantum theories and overcoming the zero-point energy singularity in QED [14–18]. Photons are oscillations of the electromagnetic vacuum. In addition, the electron–positron elementary charge is derived naturally during the photon vector potential quantization process. Hence, it has been shown that the photon vector potential is directly related to the elementary charge and mass, entailing that the electron and positron may be composed of standing electromagnetic vacuum states in specific topological configurations.

We advance here the possibility that all the elementary particles are composed of charges, which themselves are states of the electromagnetic vacuum which is at the origin of photons, establishing the mass-charge equivalence relation. It is deduced that the universe may be composed entirely of light issued from the electromagnetic vacuum. In addition, it is put ahead that gravitation is an electromagnetic effect, and it is shown that the gravitational constant depends directly on the electromagnetic vacuum density of states, entailing that it may not be universal.

1. Photons from the electromagnetic vacuum

Before addressing the electromagnetic nature of all elementary particles, it is of crucial importance to recall briefly the main features of the photon vector potential quantization, leading to the electromagnetic vacuum with the associated cosmological implications.

1.1. *Photon vector potential-energy quantum equation and the photon wave function*

The photon wave function $\vec{\Phi}_{k(L,R)}(\vec{r},t)$ for a free k-mode photon with angular frequency ω_k and circular polarization left (L) or right (R), $\lambda = (L,R)$ corresponding, respectively, to spin $\pm\hbar$, is expressed as follows [19, 20]:

$$\vec{\Phi}_{k(L,R)}(\vec{r},t) = \left(\frac{\varepsilon_0\omega_k}{\hbar}\right)^{1/2} [\alpha_{0k}(\omega_k)(\hat{\varepsilon}_{k(L,R)}e^{i(\vec{k}\cdot\vec{r}-\omega_k t+\theta)} + \hat{\varepsilon}^*_{k(L,R)}e^{-i(\vec{k}\cdot\vec{r}-\omega_k t+\theta)})] \tag{1}$$

where $\hat{\varepsilon}_{k(L,R)}$ is the circular polarization complex unit vector, $\vec{k}$ is the wave vector with amplitude $|\vec{k}| = 2\pi/\lambda_k$, λ_k is the wavelength of the mode k and θ is a phase parameter.

It satisfies Maxwell's propagation equation in vacuum

$$\vec{\nabla}_k^2\vec{\Phi}_{k(L,R)}(\vec{r},t) - \frac{1}{c^2}\frac{\partial^2}{\partial t^2}\vec{\Phi}_{k(L,R)}(\vec{r},t) = 0 \tag{2}$$

Helmholtz equation

$$\vec{\nabla}_k^2\vec{\Phi}_{k(L,R)}(\vec{r},t) + k^2\vec{\Phi}_{k(L,R)}(\vec{r},t) = 0 \tag{3}$$

and the photon *vector potential–energy* quantum equation

$$i\begin{pmatrix}\xi\\ \hbar\end{pmatrix}\frac{\partial}{\partial t}\vec{\Phi}_{k(L,R)}(\vec{r},t) = \begin{pmatrix}\tilde{\alpha}_{0k}\\ \tilde{H}_k\end{pmatrix}\vec{\Phi}_{k(L,R)}(\vec{r},t) \tag{4}$$

expressing both the first (energy) and second (vector potential) quantization of the electromagnetic field corresponding to k-mode photons with energy $E_k(\omega_k) = \hbar\omega_k$ and vector potential amplitude $\alpha_{0k}(\omega_k) = \xi\omega_k$.

The operators involved in Eq. (4) are written as

$$\begin{pmatrix} \tilde{\alpha}_{0k} \\ \tilde{H}_k \end{pmatrix} = -i \begin{pmatrix} \xi \\ \hbar \end{pmatrix} c\vec{\nabla}_k \tag{5}$$

where $\vec{\nabla}_k$ merely means that the gradient acts upon the mode k.

The vector potential quantization constant is

$$\xi = \frac{\hbar}{4\pi ec} \tag{6}$$

with $\hbar$ Planck's reduced constant ($\hbar = h/2\pi$), c the speed of light in vacuum and e the electron–positron elementary charge. Note that ξ can be negative or positive [14–16].

The function $\vec{\Phi}_{k(L,R)}(\vec{r},t)$ [21] issues from the photon vector potential function $\vec{\alpha}_{k(L,R)}(\vec{r},t)$ with the quantized amplitude $\alpha_{0k} = \xi\omega_k$:

$$\begin{aligned} \vec{\alpha}_{k(L,R)}(\vec{r},t) &= \xi\omega_k(\hat{\varepsilon}_{k(L,R)}e^{i(\vec{k}\cdot\vec{r}-\omega_k t+\theta)} + \hat{\varepsilon}^*_{k(L,R)}e^{-i(\vec{k}\cdot\vec{r}-\omega_k t+\theta)}) \\ &= \omega_k \vec{\Xi}_{k(L,R)}(\vec{r},t) \end{aligned} \tag{7}$$

Overcoming the point photon concept, $\vec{\Phi}_{k(L,R)}(\vec{r},t)$ is normalized with respect to the photon quantization volume V_k, an intrinsic topological property issued from both the density of states theory and the energy normalization process of a plane electromagnetic wave to a single photon:

$$V_k = \left(\frac{\hbar}{2\varepsilon_0\xi^2}\right)\omega_k^{-3} = 4\alpha\lambda_k^3 \tag{8}$$

so that the normalization condition is satisfied

$$\int_{V_k} |\vec{\Phi}_{k(L,R)}(\vec{r},t)|^2 d^3r' = 1 \tag{9}$$

Obviously, according to Eqs. (1) and (7), the photon wave function $\vec{\Phi}_{k(L,R)}(\vec{r},t)$ depends fundamentally on the main function of the vector potential $\Xi_{k(L,R)}(\vec{r},t)$:

$$\vec{\Phi}_{k(L,R)}(\vec{r},t) = \left(\frac{\varepsilon_0\omega_k^3}{\hbar}\right)^{1/2} \vec{\Xi}_{k(L,R)}(\vec{r},t) \tag{10}$$

It is straightforward to show that the mean values of the relativistic massless particle Hamiltonian $\tilde{H}_k = -i\hbar c\vec{\nabla}_k$ and the momentum operator $\vec{p}_k = -i\hbar\vec{\nabla}_k$ in the single photon state, respecting the quantization volume V_k, give the single photon energy and momentum, respectively:

$$\begin{aligned} &|\langle\Phi_{k(L,R)}(\vec{r},t)|\tilde{H}_k|\Phi_{k(L,R)}(\vec{r},t)\rangle_{V_k}| \\ &\quad = |\langle\Phi_{k(L,R)}(\vec{r},t)| - i\hbar c\vec{\nabla}_k|\Phi_{k(L,R)}(\vec{r},t)\rangle_{V_k}| = \hbar\omega_k \end{aligned} \tag{11}$$

$$\begin{aligned} &|\langle\Phi_{k(L,R)}(\vec{r},t)|\vec{p}_k|\Phi_{k(L,R)}(\vec{r},t)\rangle_{V_k}| \\ &\quad = |\langle\Phi_{k(L,R)}(\vec{r},t)| - i\hbar\vec{\nabla}_k|\Phi_{k(L,R)}(\vec{r},t)\rangle_{V_k}| = \frac{\hbar\omega_k}{c} \end{aligned} \tag{12}$$

The precession of the quantized vector potential around the propagation axis at the angular frequency ω_k with circular polarization gives birth to orthogonal electric $\vec{\varepsilon}_k$ and magnetic $\vec{\beta}_k$ fields whose amplitudes are proportional to $\xi\omega_k^2$, and the corresponding magnitudes, i.e., the square root of the sum of the squares of the components, are obtained directly from Maxwell's equations [19, 20]:

$$\begin{aligned} \|\vec{\varepsilon}_k\| &= \|\partial\vec{\alpha}_{k(L,R)}(\vec{r},t)/\partial t\| = \sqrt{2}|\xi|\omega_k^2; \\ \|\vec{\beta}_k\| &= \|\vec{\nabla}\times\vec{\alpha}_{k(L,R)}(\vec{r},t)\| = \sqrt{2\varepsilon_0\mu_0}|\xi|\omega_k^2 \end{aligned} \tag{13}$$

with μ_0 the vacuum magnetic permeability.

In addition, the energy density for a single photon state as a pointless particle is readily obtained and is identical in both classical and quantum representations:

$$\begin{aligned} W_k &= |\Phi_{k(L,R)}(\vec{r},t)|^2|\langle\Phi_{k(L,R)}(\vec{r},t)|\tilde{H}_k|\Phi_{k(L,R)}(\vec{r},t)\rangle_{V_k}| \\ &= \frac{1}{2}\left(\varepsilon_0\|\vec{\varepsilon}_k\|^2 + \frac{1}{\mu_0}\|\vec{\beta}_k\|^2\right) = \frac{\hbar\omega_k}{V_k} = 2\varepsilon_0\xi^2\omega_k^4 \end{aligned} \tag{14}$$

showing the self-consistency of the developed formalism.

1.2. *The electromagnetic vacuum*

We recall that the zero-point energy in QED issues from the harmonic oscillator Hamiltonian and represents a constant. Therefore, it commutes with all quantum mechanics Hermitian operators corresponding to observables and consequently is not responsible for the vacuum effects, such as the Lamb shift and the spontaneous emission. In fact, the last effects are calculated in QED by using the photon creation and annihilation operators [22]. In addition, it has been demonstrated that the Casimir effect is due to Lorentz's forces of source fields and not the zero-point energy [23–26]. Finally, the density of the zero-point energy conflicts with recent well-validated astrophysical observations by 120 orders of magnitude, leading to the well-known vacuum catastrophe riddle, a real cosmological problem [27–30].

In fact, the zero-point energy of the electromagnetic field issues from the fundamental mathematical ambiguity consisting of replacing commuting classical canonical variables of position and momentum by the non-commuting corresponding quantum mechanics operators [12, 15]. Consequently, the zero-point energy cannot represent a real physical state, and the question arises of how to complement the normal ordering Hamiltonian of the electromagnetic field with a realistic vacuum description.

For that purpose, let us analyze what happens to the photon functions at zero frequency. For $\omega_k = 2\pi c/\lambda_k \to 0$, i.e., for $\lambda_k \to \infty$, the photon energy, vector potential, electric and magnetic fields vanish. However, the final state does not correspond to a perfectly empty space since the fundamental function $\vec{\Xi}_{k(L,R)}(\vec{r}, t)$ of the vector potential (7) still subsists and is expressed in both classical and quantum formalisms [16–18, 31]:

$$\vec{\Xi}_{0(L,R)} = \xi[\hat{\varepsilon}_{(L,R)} e^{i\theta} + \hat{\varepsilon}^*_{(L,R)} e^{-i\theta}] \tag{15a}$$

$$\tilde{\Xi}_{0(L,R)} = \xi[\hat{\varepsilon}_{(L,R)} a_{k(L,R)} e^{i\theta} + \hat{\varepsilon}^*_{(L,R)} a^+_{k(L,R)} e^{-i\theta}] \tag{15b}$$

where in the quantum expression (15b), we have used the creation $a^+_{k(L,R)}$ and annihilation $a_{k(L,R)}$ non-Hermitian operators, respectively, for a single k-mode photon with circular polarization.

The expressions (15) represent the electromagnetic field ground state corresponding to the electromagnetic vacuum, a zero-energy field with electric potential nature filling all of the space.

For the first time, the classical electromagnetic theory is endowed with a vacuum state and QED with a vacuum representation depending on the photon creation and annihilation operators, complementing the normal ordering Hamiltonian.

Obviously, the phase parameter θ in the expressions (15) may take any value, and consequently, the electromagnetic vacuum is composed of all possible states $\Xi_{k(L,R)}(\vec{r},t)$:

$$\vec{\Xi}_{k(L,R)}(\vec{r},t) = \xi(\hat{\varepsilon}_{k(L,R)}e^{i(\vec{k}\cdot\vec{r}-\omega_k t+\phi)} + \hat{\varepsilon}^*_{k(L,R)}e^{-i(\vec{k}\cdot\vec{r}-\omega_k t+\phi)}) \quad \text{(16a)}$$

$$\tilde{\Xi}_{k(L,R)}(\vec{r},t) = \xi(\hat{\varepsilon}_{k(L,R)}a_{k(R,L)}e^{i(\vec{k}\cdot\vec{r}-\omega_k t+\phi)} + \hat{\varepsilon}^*_{k(L,R)}a^+_{k(L,R)}e^{-i(\vec{k}\cdot\vec{r}-\omega_k t+\phi)}) \quad \text{(16b)}$$

The electromagnetic vacuum states $\Xi_{k(L,R)}(\vec{r},t)$ are called *kenons* (from $\kappa\varepsilon\nu o$ = vacuum) and involve all the characteristic physical parameters for any k-mode of the electromagnetic field, i.e., wave vector, frequency and spin (helicity, corresponding to circular polarization).

It is extremely important to note that the *kenons* result naturally from the quantized vector potential without stating any postulates or advancing any hypothesis.

From the fundamental photon *vector potential–energy* quantum equation (4), we obtain the equation governing the *kenons* [31]:

$$i\frac{\partial}{\partial t}\begin{bmatrix}\vec{\Xi}_{k(L,R)}(\vec{r},t)\\ \tilde{\Xi}_{k(L,R)}(\vec{r},t)\end{bmatrix} = \tilde{\omega}_k\begin{bmatrix}\vec{\Xi}_{k(L,R)}(\vec{r},t)\\ \tilde{\Xi}_{k(L,R)}(\vec{r},t)\end{bmatrix} = \begin{bmatrix}\vec{\alpha}_{k(L,R)}(\vec{r},t)\\ \tilde{\alpha}_{k(L,R)}(\vec{r},t)\end{bmatrix} \quad (17)$$

where the angular frequency operator acting upon each mode k is $\tilde{\omega}_k = -ic\vec{\nabla}_k$.

The physical interpretation of Eq. (17) is that real photons with vector potential $\vec{\alpha}_{k(L,R)}(\vec{r},t)$ are generated by the action of the angular frequency operator $\tilde{\omega}_k$ upon the *kenons.* In other words, photons are oscillations (precessions) of the electromagnetic vacuum states.

Hence, the electromagnetic vacuum is composed of an infinite sea of *kenons* involving all modes k with circular polarizations. An interaction Hamiltonian between *kenons* and electrons can be readily defined as

$$H_{\text{int}} = \frac{\hbar e}{m_e c}\tilde{\Xi}_{k(L,R)}\tilde{\omega}_k = -i\hbar\frac{e}{m_e}\tilde{\Xi}_{k(L,R)}\vec{\nabla}_k \tag{18}$$

characterizing the spontaneous emission effect, following which a *kenon* state is transformed to a real photon under the action of the angular frequency operator.

As for the Lamb shift, it is estimated exactly with the well-known method in QED since the calculation is based on the photon creation operator $a_{k\lambda}^{+}$.

In addition, it is straightforward to demonstrate that every particle accelerated in the electromagnetic vacuum with an acceleration γ experiences the Fulling–Davies–Unruh temperature [18]:

$$T_H = \frac{\hbar}{2\pi c k_B}|\vec{\gamma}| \tag{19}$$

where k_B is Boltzmann's constant.

Finally, it has been shown that the fluctuations of the electromagnetic vacuum yield an energy density $\sim 10^{-10}\,\text{J}\,\text{m}^{-3}$, compatible with the astrophysical observations [30]. This result stems from the fact that the photon electric field magnitude is proportional to the square of the frequency [17, 20, 32].

Thus, the electromagnetic vacuum (16) is expressed in both classical and quantum formalisms, complementing the quantum representation by associating a zero-energy vacuum field to the normal ordering Hamiltonian, overcoming the zero-point energy singularity.

2. The elementary charge from the electromagnetic vacuum

We will show now that the elementary electron–positron charge is directly related to the photon quantized vector potential.

The photon volume V_k expands longitudinally along the propagation axis and corresponds to the volume in which the precession of the vector potential quantized amplitude takes place. We may now consider a closed loop in which the photon, i.e., the kenon state, is curved. The mean electric field can be expressed by considering the photon electric field amplitude (13), weighted by the ratio of the total volume in all directions ($4\pi V_k$) to the characteristic volume λ_k^3

$$\langle \varepsilon \rangle \simeq \frac{4\pi V_k}{\lambda_k^3} |\vec{\varepsilon}_k| = 16\pi\alpha\xi\omega_k^2 \tag{20}$$

On the other hand, according to Coulomb's law, the mean electric field created by a point charge at the center of the configuration is written as

$$\langle \varepsilon \rangle \simeq \frac{q}{4\pi\varepsilon_0 r_0^2} \tag{21}$$

where $r_0 = \lambda_k/4\pi$ is the mean distance value in this topological configuration.

Equating the last equations and using $\varepsilon_0\mu_0 c^2 = 1$, we deduce the charge due to this standing photon, i.e., kenon, state:

$$q = (4\pi)^2 \alpha \frac{\xi}{\mu_0} \simeq \pm 1.602\,10^{-19}\,\mathrm{C} \tag{22}$$

which is that of the electron–positron charge e.

It is extremely important to underline here that the same exact charge issues naturally from the energy normalization of a plane electromagnetic wave over a wavelength to a single photon energy [16, 19, 33].

Using the classical electromagnetic theory formalism, the spin is written as

$$\begin{aligned} \vec{S} = \int_{V_k} \varepsilon_0 \vec{r}_0 \times (\vec{\varepsilon}_{k\lambda} \times \vec{\beta}_{k\lambda}) d^3 r &= \pm\varepsilon_0 (c/2\omega_k)(\sqrt{2}\omega_k \alpha_{0k}) \\ \times (\sqrt{2}\omega_k \alpha_{0k}/c) V_k (\hat{r}_0 \times \hat{k}) &= \pm\frac{\hbar}{2}(\hat{r}_0 \times \hat{k}) \end{aligned} \tag{23}$$

with $\hat{r}_0, \hat{k}$ unit vectors.

Furthermore, it can be shown [18, 31] that the electron–positron mass m_{e^-,e^+} depends directly on the vacuum constant ξ and the elementary charge e:

$$m_{e^-,e^+} = 2\pi c\, e^2 \frac{\xi}{\mu_B} \tag{24}$$

where $\mu_B = 9.274\,10^{-24}\,\mathrm{J\,T^{-1}}$ is the Bohr magneton.

Consequently, the electron–positron charge, spin and mass are directly related to the electromagnetic vacuum. As advanced previously [34], it is quite plausible physically for the electrons and positrons to be standing states of photons, i.e., of *kenons*. Pairs of electrons–positrons can be created spontaneously and annihilate in the *kenon* sea. Hence, it appears clearly that the *kenon* fluctuations may be at the origin of the vacuum polarization. With the same token, photons with positive and negative vector potential amplitudes are spontaneously created and annihilated. The pic of the residual electromagnetic vacuum spontaneous emission should lie in the THz frequencies, and consequently, the James Webb Space Telescope might give an answer to that issue [30].

Finally, the vacuum physical characteristics, i.e., the electric permittivity ε_0 and magnetic permeability μ_0, are expressed through the fundamental physical constants $\alpha, \xi, \hbar$ and the elementary charge e, showing that they are intrinsic properties of the *kenons*:

$$\mu_0 = (4\pi)^2 \alpha \frac{\xi}{e}; \quad \varepsilon_0 = \frac{\xi e^3}{\alpha \hbar^2} \tag{25}$$

Note that the signs of ξ and e are identical, and consequently, ε_0 and μ_0 are positive quantities. It is straightforward to verify that the last expressions give directly the velocity of light in vacuum:

$$\varepsilon_0 \mu_0 = (4\pi e \xi / \hbar)^2 = 1/c^2 \tag{26}$$

Thus, from the electromagnetic vacuum, the *kenon* sea, emerge the fundamental intrinsic vacuum properties as well as the vacuum polarization.

3. Elementary particles from elementary charges: The mass-charge equivalence

From a historical point of view, we recall that Planck's constant $\hbar$ was initially related experimentally to the energy of light quanta, considered as fundamental parts of the electromagnetic waves. However, although the characteristic physical origin of Planck's constant lies in the quantization of the electromagnetic field, it is quite intriguing to realize that the same constant is used in quantum physics for the description of all the other particles. This particular fact probably witnesses the inherent electromagnetic nature of the elementary particles.

In fact, the mass m_i of any particle–antiparticle i, other than the electron–positron, is written through the electromagnetic vacuum constant ξ and the electron–positron charge e [18, 31]:

$$m_i = 2\pi c e^2 \frac{\xi}{\mu_i} \tag{27}$$

where μ_i is the magneton of the particle i.

Note that from Eq. (22), e is the electron charge for $\xi < 0$, while it corresponds to the positron charge for $\xi > 0$. Using the same sign convention and on the condition that magnetons are positive quantities, the relation (27) implies that the masses of particles and antiparticles bear naturally opposite signs. This is in agreement with previous studies on matter and antimatter [35].

We also recall that from the very first experiments, the electron charge was considered conventionally as negative, and consequently, the particle mass also appears here as negative and those of the antiparticles as positive.

An approximate formalism for μ_i related to the Bohr magneton is written as

$$\mu_i \simeq \left(\frac{16\alpha}{n_i}\right)\mu_B \tag{28}$$

where α is the fine structure constant and n_i is simply a positive integer [31].

Using the relations (6) and (28), Eq. (27) becomes

$$m_i \simeq n_i \sigma_\xi \xi = n_i \sigma_e e \tag{29}$$

with the constants $\sigma_\xi = \pi c e^2 / 8\alpha\mu_B$ and $\sigma_e = \hbar/32\alpha\mu_B$.

The last relation signifies that the masses of all particles–antiparticles derive from the electromagnetic vacuum and are multiples of the vacuum constant ξ or multiples of the elementary charge e.

Hence, (29) expresses both the electromagnetic vacuum origin of the mass and the mass–charge equivalence.

Table 1. Elementary particle masses calculated using the relation (30) and comparison to the experimental values.

n_i	m_i calculated (MeV c^{-2})	m_i experimental (MeV c^{-2})	$\lvert\delta(\%)\rvert$	Particle
24	105.0	105.65	0.57	Muon (lepton), μ^-
31	135,6	134.97	0.46	Pion (meson), π^0
32	140.0	139.57	0.30	Pion (meson), π^+, π^-
113	494.5	493.68	0.17	Kaon (meson), K^+, K^-
114	498,9	497.70	0.24	Kaon (meson), K_s^0, K_L^0
125	547.1	547.75	0.1	Eta (meson), η^0
177	774.7	775.4	0.11	Rho (meson), ρ^0, ρ^+, ρ^-
179	783.4	782.65	0.10	Omega (meson), ω
204	892.8	891.66	0.08	Kaon (meson), K^{*+}, K^{*-}
205	897.2	896.0	0.12	Kaon (meson), K^{*0}
214	936.6	938.27	0.17	Proton (baryon), p^+
215	940.9	939.56	0.15	Neutron (baryon), n^0
219	958.4	957.66	0.07	Eta prime (meson), η'
233	1019.7	1019.44	0.01	Phi (meson), ϕ
255	1116.0	1115.68	0.03	Lambda (baryon), Λ^0
271	1186.0	1189.37	0.28	Sigma (baryon), Σ^+
272	1190.4	1192.6	0.18	Sigma (baryon), Σ^0
274	1199.2	1197.45	0.14	Sigma (baryon), Σ^-
282	1234.2	1232 ± 2	0.17	Delta (baryon), $\Delta^-, \Delta^0, \Delta^+, \Delta^{++}$
300	1313.0	1314.8	0.13	Xi (baryon), Ξ^0
302	1321.7	1321.7	0.00	Xi (baryon), Ξ^-
316	1383.0	1382.8; 1383.7	0.00	Sigma (baryon), Σ^{*+}, Σ^{*0}

(Continued)

Table 1. *(Continued.)*

n_i	m_i calculated (MeV c^{-2})	m_i experimental (MeV c^{-2})	$\|\delta(\%)\|$	Particle
317	1387.4	1387.2	0.00	Sigma (baryon), Σ^{*-}
350	1531.8	1531.8	0.00	Xi (baryon), Ξ^0 resonance
351	1536.2	1535.0	0.07	Xi (baryon), Ξ^- resonance
382	1671.9	1672.45	0.03	Omega (baryon), Ω^-
406	1776.9	1777.0	0.00	Tau (lepton), τ^-
426	1864.5	1864.8	0.01	D Meson, D^0
450	1969.5	1968.4	0.05	Ds Meson, D_s^+
458	2004.5	2006.97	0.12	D Meson, D^{+0}
459	2008.9	2010.27	0.06	D Meson, D^{++}
483	2113.9	2112.3	0.07	Strange D (meson), D_s^{*+}
522	2284.6	2286.46	0.08	Lambda c (baryon), Λ_c^+
560	2451.0	2452.9; 2453.7	0.08	Sigma c (baryon), Σ_c^+, Σ_c^0
561	2455.3	2453.9	0.05	Sigma c (baryon), Σc^{++}
564	2468.4	2467.9	0.02	Xi c (baryon), Ξ_c^+, Ξ_c^0
565	2472.8	2471.0	0.07	Xi c (baryon), Ξ_c^+, Ξ_c^0
588	2573.5	2575.7	0.08	Xi c (baryon), Ξ_c^+ resonance
589	2577.9	2578.0	0.00	Xi c (baryon), Ξ_c^0 resonance
616	2696.0	2697.5	0.06	Omega c (baryon), Ω_c^0
632	2766.0	2765.9	0.00	Omega charmed (baryon), Ω_c^{*0}
681	2980.5	2980.3	0.00	Charmed Eta (meson), η_c
707	3094.3	3096.9	0.08	J/ψ (meson)
804	3518.9	3518.9	0.00	Xi cc (baryon), Ξ_{cc}^+

(Continued)

Table 1. (*Continued.*)

n_i	m_i calculated ($\mathrm{MeV\,c^{-2}}$)	m_i experimental ($\mathrm{MeV\,c^{-2}}$)	$\lvert\delta(\%)\rvert$	Particle
827	3619.5	3621.4	0.05	Xi cc (baryon), Ξ_{cc}^{++}
1,206	5278.3	5279.34; 5279.65	0.02	B Meson, B^+, B^0
1,217	5326.4	5325.1	0.02	B* Meson, B^{*+}, B^{*0}
1,226	5365.8	5366.3	0.00	Bs Meson, B_s
1,284	5619.7	5620.2	0.00	Lambda 0s (baryon), Λ_b^0
1,327	5807.9	5810	0.04	Sigma b (baryon), Σ_b^+
1,329	5816.6	5815	0.02	Sigma b (baryon), Σ_b^-
1,332	5829.8	5830.3	0.00	Sigma bottom (baryon), Σ_b^{*+}
1,333	5834.1	5834.7	0.00	Sigma bottom (baryon), Σ_b^{*-}
1,356	5934.8	5935.0	0.00	Xi' b (baryon), Ξ_b'
1,360	5952.3	5952.3	0.00	Xi*0 (baryon), Ξ_b^{*0}
1,361	5956.6	5955.3	0.02	Xi*b (baryon), Ξ_b^*
1,381	6044.2	6046.0	0.02	Omega b (baryon), Ω_b
1,423	6228.0	6226.9	0.01	Xi b (baryon), Ξ_b
1,434	6276.1	6276 ± 4	0.00	Charmed B Meson, B_c^+
2,125	9300.4	9300 ± 40	0.00	Eta bottom (meson), η_b
2,161	9458.0	9460.3	0.02	Ypsilon (meson), Y

The numerical application of (29) is written simply as [31]

$$m_i \simeq n_i 4.3767\,\mathrm{MeV/c^2} \tag{30}$$

This extremely simple relation reproduces the masses of all known elementary particles with a precision less than 0.5%, hence within the experimental error limits (Table 1).

The integer characterizing the particle masses in Eq. (29) entails that the electromagnetic vacuum must be endowed with a complex structure involving specific quantum states that might be related to string theory.

In addition, quarks and antiquarks are characterized by fractional elementary charges, e/3 and 2e/3, and could also correspond to particular states of the electromagnetic vacuum. Therein, it is interesting to underline that the value $4.3767\,\mathrm{MeV/c^2}$ in the relation (30) corresponds to the mass of the down quark, which is estimated to be roughly $4.7 \pm 0.6\,\mathrm{MeV/c^2}$.

Pairs of particles–antiparticles can be created spontaneously and annihilate in vacuum due to the kenon fluctuations. However, the probability for this mechanism to occur is much weaker than that of electron–positron pairs, for which the fluctuating energy involved in the process is much lower. The residual spontaneous generation-annihilation of matter-antimatter in space by the electromagnetic vacuum fluctuations may be at the origin of vacuum polarization.

Finally, the overall conclusion of the above is that the electron–positron charge is derived from the electromagnetic vacuum and is directly related to photons through the vector potential. The particle–antiparticle masses originate from the elementary charges and their magnetic moment which are states of the electromagnetic vacuum in specific topological configurations.

4. The gravitational constant and the electromagnetic vacuum

It is well known that Planck's length $l_P = 1.616\,10^{-35}$ m corresponds physically to the shorter possible wavelength of a single photon. Consequently, it characterizes the "granularity" of the electromagnetic vacuum. Considering l_P, it has been shown that the gravitational constant G is expressed exactly through the electromagnetic vacuum constants ξ, α and ε_0 [17, 18]:

$$G = \frac{1}{4\pi\varepsilon_0\alpha}\left(\frac{l_p}{4\pi\xi}\right)^2 \tag{31}$$

Assuming that Newton's gravitational law is valid at the elementary particle scale, the equivalence between Coulomb's electrostatic law and Newton's gravitational law is straightforward [31]:

$$U_{\text{Newton}} = G\frac{m_i m_j}{r_{ij}} = \frac{1}{4\pi\varepsilon_0}\frac{e_i e_j}{r_{ij}}\eta_{ij} = U_{\text{Coulomb}} \tag{32}$$

where $\eta_{ij} = \frac{\pi\hbar c l_P^2}{\mu_0\mu_i\mu_j}$ is a dimensionless parameter, depending on the magnetons μ_i and μ_j of the interacting particles.

The gravitational potential between a large number of particles is expressed uniquely through the QED constants $\hbar$, e and the magnetons μ_i, μ_j, showing the electromagnetic nature of the gravitational interaction:

$$U = \frac{\hbar^2}{4}G\sum_{i,j(i<j)}\frac{e_i e_j}{r_{ij}\mu_i\mu_j} \tag{33}$$

Now, the factor $4\pi\xi/l_p$ in (32) corresponds to the linear density of the vacuum states with respect to the granularity of the electromagnetic vacuum in 4π steradians.

Consequently, a fundamental concept arises straightforwardly related to the gravitational constant dependence on the electromagnetic vacuum density. In fact, the relation (31) is written in a more general expression

$$G = \frac{1}{4\pi\varepsilon_0\alpha}\frac{1}{\rho(\xi)^2} \tag{34}$$

where $\rho(\xi)$ is the density of states of the electromagnetic vacuum.

We deduce that the gravitational constant G is derived from the electromagnetic vacuum, and according to the variation of the *kenon* density in space, it may not be a universal constant. This could provide a direct explanation of the gravitational anomalies observed on quite a large number of galaxies, which conflict with Newtonian dynamics and general relativity [36]. On the other hand, following many detailed studies over 193 high-quality disk galaxies, it is also well established today with a high degree of statistical accuracy that the modified Newtonian dynamics models are not valid [37]. Consequently, the local variation of the gravitational constant due to the electromagnetic vacuum density could give an explanation for those astrophysical anomalies. Furthermore, the radiation pressure of the electromagnetic vacuum might be at the origin of gravitation (*Electromagnetic Push Gravity*) and would be interesting to be investigated experimentally [31, 38]. In addition, the opposite mass signs corresponding to matter and antimatter following the relation (29) may easily interpret Hubble's law as well as cosmic acceleration in agreement with recent astrophysical simulations [39].

Finally, the Breit–Wheeler effect [40], according to which high-energy photon collisions give birth to electron–positron pairs, has been confirmed [41, 42], while advanced experiments [43, 44] managed to create a high number of electron–positron pairs using γ photon collisions. Consequently, the experimental evidence has demonstrated that matter-antimatter can be created from photons, opening new perspectives for further investigations in order to understand whether the universe is entirely composed of light.

5. Conclusion

The electromagnetic vacuum is a real universal field composed of *kenon* states with an electric potential nature. It has both classical and quantum representations and is at the origin of the fundamental intrinsic vacuum properties, such as the electric permittivity, magnetic permeability and vacuum polarization. The electromagnetic waves, photons in quantum theory, are simply *kenon* oscillations.

Electrons and positrons are bound topological configuration states of photons, i.e., of *kenons*. The elementary charge results precisely from the *kenons* and is proportional to the electromagnetic vacuum amplitude ξ.

The masses of all the elementary particles can be expressed as multiples of the vacuum quantization amplitude ξ and thus of the elementary charge e. It is concluded that matter and antimatter in the universe are composed of *kenons*, i.e., of light.

The electromagnetic nature of particles implies the electromagnetic nature of gravitation and yields the equivalence of Newton's gravitational law to Coulomb's electrostatic law. Considering Planck's length as the fundamental granularity of the electromagnetic vacuum, we have shown that the gravitational constant has an electromagnetic nature, and it's inversely proportional to the square of the *kenon* density in space and consequently might not be a universal constant.

References

[1] V. Ronchi, *The Nature of Light: An Historical Survey* (Harvard University Press, Cambridge, MA, USA, 1970).

[2] F. H. Read, *Electromagnetic Radiation* (John Wiley & Sons, New York, NY, USA, 1980).

[3] J. D. Jackson, *Classical Electrodynamics*, 3rd ed. (Wiley, New York, NY, USA, 1998).

[4] W. Ehrenberg and R. E. Siday, The refractive index in electron optics and the principles of dynamics, *Proc. Phys. Soc. London, Sect. B.* 62, 8 (1949).

[5] R. G. Chambers, Shift of an electron interference pattern by enclosed magnetic flux, *Phys. Rev. Lett.* 5, 3–5 (1960).

[6] A. Tonomura *et al.*, Observation of Aharonov-Bohm effect by electron holography, *Phys. Rev. Lett.* 48(21), 1443 (1982).

[7] N. Osakabe *et al.*, Experimental confirmation of Aharonov-Bohm effect using a toroidal magnetic field confined by a superconductor, *Phys. Rev. A.* 34, 815 (1986).

[8] R. P. Feynman, *QED: The Strange Theory of Light and Matter* (Princeton University Press, Princeton, NJ, USA, 1988).

[9] W. Heitler, *The Quantum Theory of Radiation* (Clarendon Press, Oxford, UK, 1954).

[10] M. H. Mittleman, *Introduction to the Theory of Laser-Atom Interactions* (Plenum Press, New York, NY, USA, 1982).

[11] M. Weissbluth, *Photon-Atom Interactions* (Academic Press, Boston, MA, USA, 1988).

[12] J. C. Garrison and R. Y. Chiao, *Quantum Optics* (Oxford University Press, Oxford, UK, 2008).

[13] G. Grynberg, A. Aspect and C. Fabre, *Introduction to Quantum Optics* (Cambridge University Press, Cambridge, UK, 2010).

[14] C. Meis, Vector potential quantization and the quantum vacuum, *Phys. Res. Int.* 2014, 187432 (2014).

[15] C. Meis, *Light and Vacuum*, 2nd ed. (World Scientific, Singapore, 2017).
[16] C. Meis, Quantized field of single photons, in *Single Photon Manipulation*, K. Xia, Ed. (IntechOpen, London, UK, 2019).
[17] C. Meis, The electromagnetic field ground state and the cosmological evolution, *J. Phys.: Conf. Ser.* 1141, 012072 (2018).
[18] C. Meis, Primary role of the quantum electromagnetic vacuum in gravitation and cosmology, in *Cosmology 2020 — The Current State*, M. L. Smith, Ed. (IntechOpen, London, UK, 2020).
[19] C. Meis, Photon structure and wave function from the vector potential quantization, *J. Mod. Phys.* 14, 311–329 (2023).
[20] C. Meis, Photon wave function, electric and magnetic fields: Relationship to the electromagnetic vacuum, OPTICA OPEN ID 113886 (2024).
[21] C. Meis and P. R. Dahoo, Quantized vector potential and the photon wave function, *J. Phys.: Conf. Ser.* 936, 012004 (2017).
[22] P. W. Milonni, *The Quantum Vacuum* (Academic Press, London, UK, 1994).
[23] J. Schwinger, L. L. DeRaad, Jr. and K. A. Milton, Casimir effect in dielectrics, *Ann. Phys.* 115, 1 (1978).
[24] P. W. Milonni, Casimir forces without the vacuum radiation field, *Phys. Rev. A.* 25, 1315 (1982).
[25] H. Ezawa, K. Nakamura and K. Watanabe, The Casimir force from Lorentz's approach, in *Frontiers in Quantum Physics*, S. C. Lim, R. Abd-Shukor, and K. H. Kwek, Eds. (Springer, Singapore, 1998), pp. 160–169.
[26] G. Gründler, The Casimir effect: No manifestation of zero-point energy, arXiv:1303.3790 (2013).
[27] J. A. Frieman, M. S. Turner and D. Huterer, Dark energy and the accelerating universe, *Annu. Rev. Astron. Astrophys.* 46, 385 (2008).
[28] T. Padmanabhan, Cosmological constant: The weight of the vacuum, *Phys. Rep.* 380, 235 (2003).
[29] G. R. Bengochea, G. León, E. Okon, and D. Sudarsky, Can the quantum vacuum fluctuations really solve the cosmological constant problem? *Eur. Phys. J. C.* 80, 18 (2020).
[30] C. Meis, The vacuum electromagnetic energy density: Towards a physical comprehension of the vacuum energy scale problem, *Int. J. Mod. Phys. A* 35, 2050153 (2020).
[31] C. Meis, The electromagnetic nature of gravitation and matter–antimatter antigravity: Surmise on quantum vacuum gravitation and cosmology, *J. Mod. Phys.* 13, 949–968 (2022).

[32] C. Meis and P. R. Dahoo, Vector potential quantization and the photon intrinsic electromagnetic properties: Towards nondestructive photon detection, *Int. J. Quantum Inf.* 15(8), 1740003 (2017).
[33] C. Meis, Zero-point radiation field revisited, *Phys. Essays* 12, 1 (1999).
[34] J. G. Williamson and M. B. van der Mark, Is the electron a photon with toroidal topology? *Ann. Fond. Louis de Broglie* 22, 133 (1997).
[35] M. Villata, CPT symmetry and antimatter gravity in general relativity, *Europhys. Lett.* 94, 20001 (2011).
[36] M. Fairbairn, Galactic anomalies and particle dark matter, *Symmetry* 14(4), 812 (2022).
[37] D. C. Rodrigues, V. Marra, A. Del Popolo and Z. Davari, Absence of a fundamental acceleration scale in galaxies, *Nat. Astron.* 2(6), 68–672 (2018).
[38] G. Danilatos, Novel quantitative push gravity/electricity theory poised for verification, Zenodo (2020).
[39] C. Dimopoulos *et al.*, Hubble law and acceleration curve emerges in a repulsive matter–antimatter galaxies simulations, *Astropart. Phys.* 147, 102806 (2023).
[40] G. Breit and J. A. Wheeler, Collision of two light quanta, *Phys. Rev.* 46, 1087–1091 (1934).
[41] C. Bula *et al.*, Observation of nonlinear effects in Compton scattering, *Phys. Rev. Lett.* 76, 3116–3119 (1996).
[42] C. Bamber *et al.*, Studies of nonlinear QED in collisions of 46.6 GeV electrons with intense laser pulses, *Phys. Rev. D* 60(9), 092004 (1999).
[43] O. J. Pike *et al.*, A photon–photon collider in a vacuum hohlraum, *Nat. Photonics* 8, 434 (2014).
[44] J. Adam *et al.* (STAR Collaboration), Measurement of e+e− momentum and angular distributions from linearly polarized photon collisions, *Phys. Rev. Lett.* 127, 052302 (2021).

Conclusion

Here, the results pertaining to the themes covered in Part II are summarized, listing the following new concepts that have been introduced:

- Photons, electrons and positrons are strongly related to the electromagnetic vacuum.
  The electron and positron charge and mass issue from the electromagnetic vacuum and depend directly on the vacuum constant ξ, which is also the quantization constant of the photon vector potential
  $$e = (4\pi)^2 \alpha \frac{\xi}{\mu_0} \simeq \pm 1.602\ 10^{-19}\mathrm{C}$$
  $$m_{e^-,e^+} = 2\pi c\, e^2 \frac{\xi}{\mu_B} \simeq 0.51\ \mathrm{MeV/c^2}$$
- The electromagnetic vacuum energy density and the cosmological constant: The mean energy density in space issued from the electromagnetic vacuum fluctuations is compatible with the astrophysical observations on the dark energy:
  $$\rho_{\text{vacuum}} \sim \frac{\hbar}{\pi^2 c^3} \int_0^\infty p_0 e^{-\omega^2/\omega_m^2} \omega^3 d\omega = p_0 \frac{\hbar\, \omega_m^4}{2\,\pi^2 c^3}$$
  with $p_0 \sim 1$ and ω_m in the THz frequency region, for $\omega_m \sim 10^{12}\mathrm{Hz}$, the obtained vacuum energy density is $\rho_{\text{vacuum}} \sim 3 \cdot 10^{-10}\mathrm{J\,m^{-3}}$. The electromagnetic vacuum irradiates weakly in the THz spectrum.

- Mass-charge equivalence implies the electromagnetic nature of gravitation. The charge states issue from the electromagnetic vacuum. The mass effect is the result of charge states composing the particle and of their magnetic moment. The masses of all the elementary particles are proportional to the vacuum constant ξ or to the elementary charge e times an integer n_i and obtained with a precision less than 0.5%:

$$m_i \simeq n_i \sigma_\xi \xi = n_i \sigma_e e$$

with the constants $\sigma_\xi = \pi c e^2/8\alpha\mu_B$ and $\sigma_e = \hbar/32\alpha\mu_B$
The relationship to string theory should be investigated.

- Coulomb's electrostatic and Newton's gravitational laws' equivalence

$$U_{\text{Newton}} = G\frac{m_i m_j}{r_{ij}} = \frac{1}{4\pi\varepsilon_0}\frac{e_i e_j}{r_{ij}}\eta_{ij} = U_{\text{Coulomb}}$$

where η_{ij} is a dimensionless parameter depending on the magnetons μ_i and μ_j of the interacting particles.

- The gravitational constant depends on the electromagnetic vacuum density

$$G = \frac{1}{4\pi\varepsilon_0\alpha}\frac{1}{\rho(\xi)^2}$$

where $\rho(\xi)$ is the density of states of the electromagnetic vacuum and α the fine structure constant.
Consequently, G might not be a universal constant.

Epilogue

"The more you know the more you realize that you don't know."

— Aristotle

Since the introduction of the photon concept, roughly a century ago, this physical entity has remained enigmatic till today, constantly puzzling the scientific community. Despite the impressive technological progress during the last decades, permitting us to manipulate single photons thanks to parametric down-conversion, we are still unable today to provide a clear mathematical description of a single photon. Obviously, the statement expressed by Albert Einstein 70 years ago and quoted in the introduction of Chapter 7 is valid even today.

The collection of the selected papers presented in this book represents chronologically the efforts of the author to understand the real nature of photons. This study revealed that some fundamental concepts of the QED theory, which remained static for almost a century, have to be reappraised, hence a detailed synthesis is presented in Chapter 7.

The vector potential amplitude of a free photon and the corresponding quantum mechanics operators are now clearly established, independent of any external cavity parameter. Using the photon vector potential amplitude operator, it is straightforward to establish the photon quantum equation coupling the first (energy) and second (vector potential) quantization of the electromagnetic field.

A single photon is not a point and occupies a minimum quantization volume, issued from both the density of states theory and the energy normalization procedure. Thus, a photon wave function is now readily established based on the quantization of the vector potential function and normalized with respect to the quantization volume, satisfying the photon quantum equation.

During the quantization procedure of the vector potential at a single photon level, the electron–positron elementary charge as well as the electromagnetic vacuum appear naturally. The single photon electric and magnetic fields amplitudes are proportional to the square of the frequency. Photons are clearly oscillations of the vacuum states. For the first time, the electron–positron charge is directly related to the photon vector potential and the electromagnetic vacuum. Photons, electrons and positrons issue from the same vacuum field and are strongly related entities. Under specific experimental conditions photons can be created from the electromagnetic vacuum.

We have shown that the cosmological constant can be calculated from the electromagnetic vacuum fluctuations, achieving a result quite compatible with the astrophysical observations. It comes out that the vacuum should spontaneously emit a weak radiation in the THz range of frequencies, which is worthwhile to be investigated experimentally. Unexpectedly, these developments may now ensure a pertinent link among QED, particle physics and cosmology.

Finally, it is important to note that without stating postulates or advancing any hypothesis, the masses of all known elementary particles appear to be integer multiples of the electromagnetic vacuum constant ξ or the elementary charge e. This may also establish a link between QED and string theory.

It comes out that all particles are "charge states", and pure neutral particles without a charge composition do not exist. Hence, the universe is composed of photons and charge states, i.e., electrons, protons and neutrons, which are all related to the electromagnetic vacuum. This reasonably entails that the gravitation should have an electromagnetic nature. The equivalence between Coulomb's electrostatic law and Newton's gravitational law is put in evidence. Furthermore, we have shown that the gravitational constant is inversely proportional to the square of the electromagnetic vacuum density of states, suggesting that it might not be a universal constant.

Consequently, the electromagnetic vacuum issued from the QED theory plays a basic role in the origin and evolution of the universe, and it is worth carrying out further experimental studies on this topic.

Index

D

E

F

G

H

I

K

L

M

N

O

P

Q

R

www.ingramcontent.com/pod-product-compliance
Lightning Source LLC
LaVergne TN
LVHW051051210926
848486LV00002B/71